D0872106

DATE DUE

The
Chemical Formulary

*Collection of Commercial Formulas
for Making Thousands of products
in Many Fields*

VOLUME XXII

Editor-in-Chief

H. BENNETT, F.A.I.C.

**Director, B. R. Laboratory
(Formula Consultants)
Miami Beach, Florida 33140**

CHEMICAL PUBLISHING COMPANY, INC.

New York

1979

© 1979

H. Bennett

ISBN 0-8206-0280-9

Printed in the United States of America

CONTRIBUTORS

Abbott, W. K.   I.C.I. Americas Inc.   Wilmington, Del.

Abbott, W. K.	I.C.I. Americas Inc.	Wilmington, Del.
Carlson, B. C.	Vanderbilt Co., R.T.	Norwalk, Ct.
Christensen, E. H.	Quaker Oats Co.	Chicago, Ill.
Conte, J. G. '	Witco Chemical Corp.	Paterson, N.J.
Cramer, Marvin B.	Goldschmidt Products Corp.	White Plains, N.Y.
Davis, T. R.	Polyesther Corp.	Southampton, N.Y.
DeBagon, S. A.	Amerchol	Edison, N.J.
DeBeers, Jr. F. M.	deBeers Lab.	Addison, Ill.
Dezeih, C. J.	Aloe Creme Labs	Ft. Lauderdale, Fla.
Doepker, M. L.	Consulting Chemist	Miami, Fla.
France, J. R.	Patco Products Co.	Kansas City, Mo.
Goode, S.	R I T A Chemical Corp.	Crystal Lake, Ill.
Hart, J. Roger	W.R. Grace Co.	Nashua, N.H.
Idson, B.	Hoffman LaRoche Inc.	Nutley, N.J.
Keene, C. N.	Monsanto Co.	St. Louis, Mo.
Leitner, G. J.	Stauffer Chemical Co.	Dobbs Ferry, N.Y.
Mahoney, W. E.	Humko Sheffield	Lyndhurst, N.J.
Meehan, P. P.	Sherwin Williams Co.	Chicago, Ill.
Neulinger, K. F.	Croda Inc.	New York, N.Y.
O'Connor, M. J.	Hexcel Fine Organics	Lodi, N.J.
Patel, S. K.	PVO International	Boonton, N.J.
Peterkofsky, Alan L.	Alcolac	Baltimore, Md.
Posselt, F. W.	GAF Corp.	New York, N.Y.
Schmolka, I. R.	BASF Wyandotte	Wyandotte, Mich.
Seldner, A.	Amerchol	Edison, N.J.
Slipiec, Roy E.	Mazer Chemicals Inc.	Gurnec, Ill.
Tramel, J. E.	Stauffer Chemical Co.	Atlanta, Ga.
Vaughn, C. D.	Kolmar Labs Inc.	Port Jervis, N.Y.

BOOKS BY H. BENNETT

The Chemical Formulary Vol. I–XXI
The Cumulative Index—The Chemical Formulary
Concise Chemical & Technical Dictionary
New Cosmetic Formulary
Chemical Specialties
Industrial Waxes, Vols. I, II
Practical Emulsions, Vols. I, II
More For Your Money
Trademarks, Chemical

TABLE OF CONTENTS

PREFACE

Chemistry, as taught in our schools and colleges, concerns chiefly synthesis, analysis, and engineering—and properly so. It is part of the right foundation for the education of the chemist.

Many a chemist entering an industry soon finds that most of the products manufactured by his concern are not synthetic or definite chemical compounds, but are mixtures, blends, or highly complex compounds of which he knows little or nothing. The literature in this field, if any, may be meager, scattered, or obsolete.

Even chemists with years of experience in one or more industries spend considerable time and effort in acquainting themselves with any new field which they may enter. Consulting chemists similarly have to solve problems brought to them from industries foreign to them. There was a definite need for an up-to-date compilation of formulae for chemical compounding and treatment. Since the fields to be covered are many and varied, an editorial board of chemists and engineers engaged in many industries was formed.

Many publications, laboratories, manufacturing firms, and individuals have been consulted to obtain the latest and best information. It is felt that the formulas given in this volume will save chemists and allied workers much time and effort.

Manufacturers and sellers of chemicals will find, in these formulae, new uses for their products. Nonchemical executives, professional men, and interested laymen will make through this volume a "speaking acquaintance" with products which they may be using, trying, or selling.

It often happens that two individuals using the same ingredients in the same formula get different results. This may be due to slight deviations in the raw materials or unfamiliarity with the intricacies of a new technique. Accordingly, repeated experiments may be necessary to get the best results. Although many of the formulas given are being used commercially, many have been taken from the literature and may be subject to various errors and omissions. This should be taken into consideration. Wherever possible, it is advisable to consult with other chemists or technical workers regarding commercial production. This will save time and money and help avoid trouble.

A formula will seldom give exactly the results which one requires. Formulas are useful as starting points from which to work out one's ideas. Also, formulas very often give us ideas which may help us in our specific problems. In a compilation of this kind, errors of omission, commission, and printing may occur. I shall be glad to receive any constructive criticism.

Included are the new F.D.A. regulations for drugs, cosmetics, and pesticides.

H. BENNETT

PREFACE TO VOLUME XXII

This new volume of the CHEMICAL FORMULARY series is a collection of new, up-to-date formulas. The only repetitious material is the introduction (Chapter I) which is used in every volume for the benefit of those who may have bought only one volume and who have no educational background or experience in chemical compounding. The simple basic formulas and compounding methods given in the introduction will serve as a guide for beginners and students. It is suggested that they read the introduction carefully and even make a few preparations described there before compounding the more intricate formulas included in the later chapters.

The list of chemicals and their suppliers has been enlarged with new trademark chemicals, so that buying the required ingredients will present no problem.

Grateful acknowledgement is made to the Contributors for their valuable suggestions and contributions.

H. BENNETT

NOTE: All the formulas in Volumes I through XXII (except in the Introduction) are different. Thus, if you do not find what you want in this volume, you may find it in one of the others.

NOTE: This book is the result of cooperation of many chemists and engineers who have given freely of their time and knowledge. It is their business to act as consultants and to give advice on technical matters for a fee. As publishers, we do not maintain a laboratory or consulting service to compete with them. Therefore, please do not ask us for advice or opinions, but consult a chemist.

ABBREVIATIONS

```
amp .......................................... ampere
amp/dm² ...................... amperes per square decimeter
amp/sq ft ....................... amperes per square foot
anhydr ...................................... anhydrous
avoir ....................................... avoirdupois
bbl ........................................... barrel
Bé ........................................... Baumé
B.P. ...................................... boiling point
°C ...................................... degrees Centigrade
cc ...................................... cubic centimeter
cd ....................................... current density
cm ......................................... centimeter
cm³ ..................................... cubic centimeter
conc ...................................... concentrated
c.p. ...................................... chemically pure
cp ......................................... centipoise
cu ft ....................................... cubic foot
cu in ....................................... cubic inch
cwt ....................................... hundredweight
d ........................................... density
dil .......................................... dilute
dm ......................................... decimeter
dm² ..................................... square decimeter
dr .......................................... dram
E ........................................... Engler
°F ...................................... degrees Fahrenheit
ffc ...................................... free from chlorine
ffpa ..................................... free from prussic acid
fl dr ....................................... fluid dram
fl oz ....................................... fluid ounce
ft pt ....................................... flash point
F.P. ...................................... freezing point
ft ........................................... foot
ft² ....................................... square foot
g ........................................... gram
gal ......................................... gallon
gr .......................................... grain
```

ABBREVIATIONS

hl . . . . hectoliter
hr . . . . hour
in . . . . .inch
kg . . . . .kilogram
l . . . . .liter
lb . . . . pound
liq . . . . .liquid
m . . . . .meter
min . . . . .minim, minute
ml . . . . .mililiter (cubic centimeter)
mm . . . . .millimeter
M.P. . . . . .melting point
N . . . . .Normal
N.F. . . . . National Formulary
oz . . . . .ounce
pH . . . . hydrogen-ion concentration
p.p.m . . . . .parts per million
pt . . . . .pint
pwt . . . . pennyweight
q.s. . . . . a quantity sufficient to make
qt . . . . quart
r.p.m. . . . . revolutions per minute
sec . . . . second
sp . . . . .spirits
Sp. Gr. . . . . .specific gravity
sq. dm. . . . . square decimeter
tech . . . . .technical
tinc. . . . . tincture
tr . . . . tincture
Tw . . . . Twaddell
U.S.P. . . . . .United States Pharmacopeia
v . . . . volt
visc. . . . . viscosity
vol . . . . .volume
wt . . . . weight

Chapter I

INTRODUCTION

The following introductory matter has been included at the suggestion of teachers of chemistry and home economics.

This section will enable anyone, with or without technical education or experience, to start making simple products without any complicated or expensive machinery. For commercial production, however, suitable equipment is necessary.

Chemical specialties are composed of pigments, gums, resins, solvents, oils, greases, fats, waxes, emulsifying agents, dyestuffs, perfumes, water, and chemicals of great diversity. To compound certain of these with some of the others requires definite and wellstudied procedures, any departure from which will inevitably result in failure. The steps for successful compounding are given with the formulas. Follow them rigorously. If the directions require that (*a*) is added to (*b*), carry this out literally, and do not reverse the order. The preparation of an emulsion is often quite as tricky as the making of mayonnaise. In making mayonnaise, you add the oil to the egg, slowly, with constant and even stirring. If you do it correctly, you get mayonnaise. If you depart from any of these details: if you add the egg to the oil, or pour the oil in too quickly, or fail to stir regularly, the result is a complete disappointment. The same disappointment may be expected if the prescribed procedure of any other formulation is violated.

The point next in importance is the scrupulous use of the proper ingredients. Substitutions are sure to result in inferior quality, if not in complete failure. Use what the formula calls for. If a cheaper product is desired, do not prepare it by substituting a cheaper ingredient for the one prescribed: use a different formula. Not infrequently, a formula will call for an ingredient which is difficult to obtain. In such cases, either reject the formula or substitute a similar substance only after a preliminary experiment demonstrates its usability. There is a limit to which this rule may reasonably be extended. In some cases, substitution of an equivalent ingredient may be made legitimately. For example, when the formula calls

1

for white wax (beeswax), yellow wax can be used, if the color of the finished product is a matter of secondary importance. Yellow beeswax can often replace white beeswax making due allowance for color, but paraffin wax will not replace beeswax, even though its light color seems to place it above yellow beeswax.

This leads to the third point: the use of good-quality ingredients, and ingredients of the correct quality. Ordinary lanolin is not the same thing as anhydrous lanolin. The replacement of one with the other, weight for weight, will give discouragingly different results. Use exactly what the formula calls for: if you are not acquainted with the substance and you are in doubt as to just what is meant, discard the formula and use one you understand. Buy your chemicals from reliable sources. Many ingredients are obtainable in a number of different grades: if the formula does not designate the grade, it is understood that the best grade is to be used. Remember that a formula and the directions can tell you only part of the story. Some skill is often required to attain success. Practice with a small batch in such cases until you are sure of your technique. Many examples can be cited. If the formula calls for steeping quince seed for 30 min in cold water, steeping for 1 hour may yield a mucilage of too thin a consistency. The originator of the formula may have used a fresher grade of seed, or his conception of what "cold" water means may be different from yours. You should have a feeling for the right degree of mucilaginousness, and if steeping the seed for 30 min fails to produce it, steep them longer until you get the right kind of mucilage. If you do not know what the right kind is, you will have to experiment until you find out. This is the reason for the recommendation to make small experimental batches until successful results are obtained. Another case is the use of dyestuffs for coloring lotions and the like. Dyes vary in strength; they are all very powerful in tinting value; it is not always easy to state in quantitative terms how much to use. You must establish the quantity by carefully adding minute quantities until you have the desired tint. Gum tragacanth is one of those products which can give much trouble. It varies widely in solubility and bodying power; the quantity listed in the formula may be entirely unsuitable for your grade of tragacanth. Therefore, correction is necessary, which can be made only after experiments with the available gum.

In short, if you are completely inexperienced, you can profit greatly by experimenting. Such products as mouthwashes, hair tonics, and astringent lotions need little or no experience, because they are, as a rule, merely mixtures of simple liquid and solid ingredients, which dissolve without difficulty and the end product is a clear solution that is ready for use when mixed. However, face creams, toothpastes, lubricating greases, wax polishes, etc., whose formulation requires relatively elaborate procedures and which must have a definte final viscosity, need some skill and not infrequently some experience.

FIGURING

Some prefer proportions expressed by weight or volume, others use percentages. In different industries and foreign countries different systems of weights and measures are used. For this reason, no one set of units could be satisfactory for everyone. Thus diverse formulas appear with different units, in accordance with their sources of origin. In some cases, parts are given instead of percentage or weight or volume. On the pages preceding the index, conversion tables of weights and measures are listed. These are used for changing from one system to another. The following examples illustrate typical units:

EXAMPLE NO. 1

Ink for Marking Glass

Glycerin	40	Ammonium Sulfate	10
Barium Sulfate	15	Oxalic Acid	8
Ammonium Bifluoride	15	Water	12

Here no units are mentioned. In this case, it is standard practice to use parts by weight throughout. Thus here we may use ounces, grams, pounds, or kilograms as desired. But if ounces are used for one item, the ounce must be the unit for all the other in the formula.

EXAMPLE NO. 2

Flexible Glue

Powdered Glue	30.90%	Glycerin	5.15%
Sorbitol (85%)	15.45%	Water	48.50%

Where no units of weight or volume, but percentages are given, forget the percentages and use the same method as given in Example No. 1.

Example No. 3

Antiseptic Ointment

Petrolatum	16 parts	Benzoic Acid	1 part
Coconut Oil	12 parts	Chlorothymol	1 part
Salicylic Acid	1 part		

The instructions given for Example No. 1 also apply to Example No. 3. In many cases, it is not wise to make up too large a quantity of a product before making a number of small batches to first master the necessary technique and also to see whether the product is suitable for the particular purpose for which it is intended. Since, in many cases, a formula may be given in proportions as made up on a factory scale, it is advisable to reduce the quantities proportionately.

Example No. 4

Neutral Cleansing Cream

Mineral Oil	80 lb	Water	90 lb
Spermaceti	30 lb	Glycerin	10 lb
Glyceryl Monostearate	24 lb	Perfume	To suit

Here, instead of pounds, ounces or even grams may be used. This formula would then read:

Mineral Oil	80 g	Water	90 g
Spermaceti	30 g	Glycerin	10 g
Glyceryl Monostearate	24 g	Perfume	To suit

Reduction in bulk may also be obtained by taking the same fractional part or portion of each ingredient in a formula. Thus in the following formula:

Example No. 5

Vinegar Face Lotion

Acetic Acid (80%)	20	Alcohol	440
Glycerin	20	Water	500
Perfume	20		

We can divide each amount by ten and then the finished bulk will be only one tenth of the original formula. Thus it becomes:

Acetic Acid (80%)	2	Alcohol	44
Glycerin	2	Water	50
Perfume	2		

APPARATUS

For most preparations, pots, pans, china, and glassware, which are used in every household, will be satisfactory. For making fine mixtures and emulsions, a malted-milk mixer or egg beater is necessary. For weighing, a small, low-priced scale should be purchased from a laboratory-supply house. For measuring fluids, glass graduates or measuring glasses may be purchased from your local druggist. Where a thermometer is necessary, a chemical thermometer should be obtained from a druggist or chemical-supply firm.

METHODS

To understand better the products which you intend to make, it is advisable that you read the complete section covering such products. You may learn different methods that may be used and also to avoid errors which many beginners are prone to make.

CONTAINERS FOR COMPOUNDING

Where discoloration or contamination is to be avoided, as in light-colored, or food and drug products, it is best to use enameled or earthenware vessels. Aluminum is also highly desirable in such cases, but it should not be used with alkalis as these dissolve and corrode aluminum.

HEATING

To avoid overheating, it is advisable to use a double boiler when temperatures below 212°F (temperature of boiling water) will suffice. If a double boiler is not at hand, any pot may be filled with water and the vessel containing the ingredients to be heated placed in the water. The pot may then be heated by any flame without fear of overheating. The water in the pot, however, should be replenished from time to time; it must not be allowed to "go dry." To get uniform higher temperatures, oil, grease, or wax is used in the outer container in place of water. Here, of course, care must be taken to stop heating when thick fumes are given off as these are inflammable. When higher uniform temperatures are necessary, molten lead may be used as a heating medium. Of course, with chemicals which melt uniformly and are nonexplosive, direct heating over an open flame is permissible, with stirring, if necessary.

Where instructions indicate working at a certain temperature, it is important to attain the proper temperature not by guesswork, but by the use of a thermometer. Deviations from indicated temperatures will usually result in spoiled preparations.

TEMPERATURE MEASUREMENT

In the United States and in Great Britain, the Fahrenheit scale of temperature is used. The temperature of boiling water is $212°$ Fahrenheit ($212°F$); the temperature of melting ice is $32°$ Fahrenheit ($32°F$).

In scientific work, and in most foreign countries, the Centigrade scale is used, on which the temperature of boiling water is $100°$ Centigrade ($100°C$) and the temperature of melting ice is $0°$ Centigrade ($0°C$).

The temperature of liquids is measured by a glass thermometer. This is inserted as deeply as possible in the liquid and is moved about until the temperature reading remains steady. It takes a short time for the glass of the thermometer to reach the temperature of the liquid. The thermometer should not be placed against the bottom or side of the container, but near the center of the liquid in the vessel. Since the glass of the thermometer bulb is very thin, it breaks easily when striking it against any hard surface. A cold thermometer should be warmed gradually (by holding it over the surface of a hot liquid) before immersion. Similarly the hot thermometer when taken out of the liquid should not be put into cold water suddenly. A sharp change in temperature will often crack the glass.

MIXING AND DISSOLVING

Ordinary dissolution (e.g., that of sugar in water) is hastened by stirring and warming. Where the ingredients are not corrosive, a clean stick, a fork, or spoon may be used as a stirring rod. These may also be used for mixing thick creams or pastes. In cases where very thorough stirring is necessary (e.g., in making mayonnaise, milky polishes, etc.), an egg beater or a malted-milk mixer is necessary.

FILTERING AND CLARIFICATION

When dirt of undissolved particles are present in a liquid, they are removed by settling or filtering. In the first procedure, the solution is allowed to stand and if the particles are heavier than the liquid they will gradually sink to the bottom. The liquid may be poured or siphoned off carefully and, in some cases, it is then sufficiently clear for use. If, however, the particles do not settle out, then they must be filtered off. If the particles are coarse they may be filtered or strained through muslin or other cloth. If they are very small, filter paper is used. Filter papers may be obtained in various degrees of fineness. Coarse filter paper filters rapidly but will not retain extremely fine particles. For fine particles, a very fine grade of filter paper should be used. In extreme cases, even this paper may not be fine enough. Then it will be necessary to add to the liquid 1 to 3% infusorial earth or magnesium carbonate. These are filter aids that clog up the pores of the filter paper and thus reduce their size and hold back undissolved material of extreme fineness. In all such filtering, it is advisable to take the first portions of the filtered liquid and pour them through the filter again as they may develop cloudiness on standing.

DECOLORIZING

The most commonly used decolorizer is decolorizing carbon. This is added to the liquid to the extent of 1 to 5% and the liquid is heated, with stirring, for ½ hour to as high a temperature as is feasible. The mixture is then allowed to stand for a while and filtered. In some cases, bleaching must be resorted to.

PULVERIZING AND GRINDING

Large masses or lumps are first broken up by wrapping in a clean cloth, placing between two boards, and pounding with a hammer. The smaller pieces are then pounded again to reduce their size. Finer grinding is done in a mortar with a pestle.

SPOILAGE AND LOSS

All containers should be closed when not in use to prevent evaporation or contamination by dust; also because, in some cases, air affects the product adversely. Many chemicals attack or corrode the metal containers in which they are kept. This is particularly true of liquids. Therefore, liquids should be transferred into glass bottles which should be as full as possible. Corks should be covered with aluminum foil (or dipped in melted paraffin wax when alkalies are present).

Glue, gums, olive oil, or other vegetable or animal products may ferment or become rancid. This produces discoloration or unpleasant odors. To avoid this, suitable antiseptics or preservatives must be used. Cleanliness is of utmost importance. All containers must be cleaned thoroughly before use to avoid various complications.

WEIGHING AND MEASURING

Since, in most cases, small quantities are to be weighed, it is necessary to get a light scale. Heavy scales should not be used for weighing small amounts as they are not accurate enough for this type of weighing.

For measuring volumes of liquids, measuring glasses or cylinders (graduates) should be used. Since this glassware cracks when heated or cooled suddenly it should not be subjected to sudden changes of temperature.

CAUTION

Some chemicals are corrosive and poisonous. In many they are labeled as such. As a precautionary measure, it is advised not to inhale them and, if smelling is absolutely necessary, only to sniff a few inches from the cork or stopper. Always work in a well-ventilated room when handling poisonous or unknown chemicals. If anything is spilled, it should be wiped off and washed away at once.

WHERE TO BUY CHEMICALS AND APPARATUS

Many chemicals and most glassware can be purchased from your druggist. A list of suppliers of all products is at the end of this book.

ADVICE

This book is the result of cooperation of many chemists and engineers who have given freely of their time and knowledge. It is their business to act as consultants and to give advice on technical matters for a fee. As publishers, we do not maintain a laboratory or consulting service to compete with them.

Please, therefore, do not ask us for advice or opinions, but confer with a chemist in your vicinity.

EXTRA READING

Keep up with new developments of materials and methods by reading technical magazines. Many technical publications are listed under references in the back of this book.

CALCULATING COSTS

Raw materials, purchased in small quantities, are naturally higher in price than when bought in large quantities. Commercial prices, as given in the trade papers and catalogs of manufacturers, are for large quantities such as barrels, drums, or sacks. For example, 1 lb. epsom salts, bought at retail, may cost 10 to 15 cents. In barrel lots its price is much lower.

Typical Costing Calculation
Formula for Beer-or Milk-Pipe Cleaner

Soda Ash	25 lb @ $0.02½ per lb =	$ 0.63
Sodium Perborate	75 lb @ 0.16 per lb =	12.00
Total	100 lb	$12.63

If 100 lb cost $12.63, 1 lb will cost $12.63 divided by 100 or about $0.126, assuming no loss.

Always weigh the amount of finished product and use this weight in calculating costs. Most compounding results in some loss of material because of spillage, sticking to apparatus, evaporation, etc. Costs of making experimental lots are always high and should not be used for figuring costs. To meet competition, it is necessary to buy in large quantities and manufacturing costs should be based on these.

ELEMENTARY PREPARATIONS

The simple formulas that follow have been selected because of their importance and because they are easy to make.

The succeeding chapters go into greater detail and give many different types and modifications of these and other recipes for home and commercial use.

Cleansing Creams

Cleansing creams, as the name implies, serve as skin cleaners. Their basic ingredients are oils and waxes which are rubbed into the skin. When wiped off, they carry off dirt and dead skin. The liquefying type cleansing cream contains no high-melting matter and melts or liquefies when rubbed on the skin. To suit different climates and likes and dislikes harder or softer products can be made.

Cleansing Cream
(Liquefying)

Liquid Petrolatum	5.5
Paraffin Wax	2.5
Petrolatum	2.0

Melt the ingredients together, with stirring, in an aluminum or enamelled dish and allow to cool. Then stir in a perfume oil. Allow to stand until it becomes hazy and then pour into jars, which should be allowed to stand undisturbed overnight.

Cold Creams

The most important facial cream is the cold cream. This type of cream contains mineral oil and wax which are emulsified in water with a small amount of borax or glyceryl monostearate, S.E. The function of a cold cream is to form a film that takes up dirt and waste tissue, which are removed when the skin is wiped thoroughly. Many modifications of this basic cream are encountered in stores. They vary in color, odor, and in claims, but, essentially, they are not more useful than this simple cream. The latest type of cold cream is the nongreasy cold cream which is of particular interest because it is nonalkaline and, therefore, nonirritating for sensitive skins.

Cold Cream

Liquid Petrolatum	52 g
White Beeswax	14 g

Heat this in a aluminum or enamelled double boiler. (The water in the outer pot should be brought to a boil.) In a separate aluminum or enamelled pot dissolve:

Borax	1 g
Water	33 cc

and bring this to a boil. Add this in a thin stream to the melted wax, while stirring vigorously in one direction only. When the temperature drops to 140°F, add 0.5 cc perfume oil and continue stirring until the temperature drops to 120°F. At this point, pour into jars, where the cream will set after a while. If a harder cream is desired, reduce the amount of liquid petrolatum. If a softer cream is wanted, increase it.

Nongreasy Cold Cream

White Paraffin Wax	1.25
Petrolatum	1.50
Glyceryl Monostearate, S.E.	2.25
Liquid Petrolatum	3.00

Heat this mixture in an aluminum or enameled double boiler. (The water in the outer pot should be boiling.) Stir until clear. To this slowly add, while stirring vigorously:

Boiling Water	10

Continue stirring until smooth and then add, with stirring, perfume oil. Pour into jars at 110 to 130°F and cover the jars as soon as possible.

Vanishing Creams

Vanishing creams are nongreasy soapy creams which have a cleansing effect. They are also used as a powder base.

Vanishing Cream

Stearic Acid	18 oz

Melt this in an aluminum or enameled double boiler. (The water in the outer pot must be boiling.) Add, in a thin stream, while stirring vigorously, the following boiling solution made in an aluminum or enameled pot:

Potassium Carbonate	¼ oz
Glycerin	6½ oz
Water	5 lb

Continue stirring until the temperature falls to 135°F, then mix in a perfume oil and stir from time to time until cold. Allow to stand overnight and stir again the next day. Pack into jars and close these tightly.

Hand Lotions

Hand lotions are usually clear or milky liquids or salves which are useful in protecting the skin from roughness and redness because of exposure to cold, hot water, soap, and other agents. Chapped hands are common. The use of a good hand lotion keeps the skin smooth, soft, and in a healthy condition. The lotion is best applied at night, rather freely, and cotton gloves may be worn to prevent soiling. During the day, it should be put on sparingly and the excess wiped off.

Hand Lotion
(Salve)

Boric Acid	1
Glycerin	6

Warm these in an aluminum or enameled dish and stir until dissolved (clear). Then allow to cool and work this liquid into the following mixture, adding only a little at a time

Lanolin	6
Petrolatum	8

To impart a pleasant odor a little perfume may be added and worked in.

Hand Lotion
(Milky liquid)

Lanolin	¼ tsp
Glyceryl Monostearate, S.E.	1 oz
Tincture of Benzoin	2 oz
Witch Hazel	25 oz

Melt the first two items together in an aluminum or enameled double boiler. If no double boiler is at hand, improvise one by placing a dish in a small pot containing boiling water. When the mixture becomes clear, remove from the double boiler and add slowly, while stirring vigorously, the tincture of benzoin and then the witch hazel. Continue stirring until cool and then put into one or two large bottles and shake vigorously. The finished lotion is a milky liquid comparable to the best hand lotions on the market sold at high prices.

Brushless Shaving Creams

Brushless or latherless shaving creams are soapy in nature and do not require lathering or water. The formula given here is of the latest type being free from alkali and nonirritating. It should be borne in mind, however, that certain beards are not softened by this type of cream and require the old fashioned lathering shaving cream.

Brushless Shaving Cream

White Mineral Oil	10
Glyceryl Monostearate, S.E.	10
Water	50

Heat the first two ingredients together in a Pyrex or enameled dish to 150°F and run in slowly, while stirring, the water which has been heated to boiling. Allow to cool to 150°F and, while stirring, add a few drops of perfume oil. Continue stirring until cold.

Mouthwashes

Mouthwashes and oral antiseptics are of practically negligible value. However, they are used because of thieir refreshing taste and slight deodorizing effect.

Mouthwash

Benzoic Acid	5/8
Tincture of Rhatany	3
Alcohol	20
Peppermint Oil	1/8

Mix together in a dry bottle until the benzoic acid is dissolved. One teaspoonful is used to a small-wine-glassful of water.

Tooth Powders

The cleansing action of tooth powders depends on their contents of soap and mild abrasives, such as precipitated chalk and magnesium carbonate. The antiseptic present is practically of no value. The flavoring ingredients mask the taste of the soap and give the mouth a pleasant aftertaste.

Tooth Powder

Magnesium Carbonate	420 g
Precipitated Chalk	565 g
Sodium Perborate	55 g
Sodium Bicarbonate	45 g
Powdered White Soap	50 g
Powdered Sugar	90 g
Wintergreen Oil	8 cc
Cinnamon Oil	2 cc
Menthol	1 g

Dissolve the last three ingredients together and then rub well into the sugar. Add the soap and perborate, mixing well. Add the chalk, with good mixing, and then the sodium bicarbonate and magnesium carbonate. Mix thoroughly and sift through a fine wire screen. Keep dry.

Foot Powders

Foot powders consist of talc or starch with or without an antiseptic or deodorizer. In the following formula the perborates liberate oxygen, when in contact with perspiration, which tends to destroy unpleasant odors. The talc acts as a lubricant and prevents friction and chafing.

Foot Powder

Sodium Perborate	3
Zinc Peroxide	2
Talc	15

Mix thoroughly in a dry container until uniform. This powder must be kept dry or it will spoil.

Liniments

Liniments usually consist of an oil and an irritant, such as methyl salicylate or turpentine. The oil acts as a solvent and tempering agent for the irritant. The irritant produces a rush of blood and warmth which is often slightly helpful.

Sore-Muscle Liniment

Olive Oil	6 fl oz
Methyl Salicylate	3 fl oz

Mix together and keep in a well-stoppered bottle. Apply externally, but do not use on chafed or cut skin.

Chest Rubs

In spite of the fact that chest rubs are practically useless, countless sufferers use them. Their action is similar to that of liniments and they differ only in that they are in the form of a salve.

Chest-Rub Salve

Yellow Petrolatum	1 lb
Paraffin Wax	1 oz
Eucalyptus Oil	2 fl oz
Menthol	½ oz
Cassia Oil	⅛ fl oz
Turpentine	½ fl oz

Melt the petrolatum and paraffin wax together in a double boiler and then add the menthol. Remove from the heat, stir, and cool a little; then mix in the oils, and turpentine. When it begins to thicken, pour into tins and cover.

Inspect Repellents

Preparations of this type may irritate sensitive skins and they will not always work.

Mosquito-Repelling Oil

Cedar Oil	2 fl oz
Citronella Oil	4 fl oz
Spirits of Camphor	8 fl oz

Mix in a dry bottle and the oil is ready for use. This preparation may be smeared on the skin as often as is necessary.

Fly Sprays

Fly sprays usually consist of deodorized kerosene, perfume, and an active insecticide. In some cases, they later recover and begin buzzing again.

Fly Spray

Deodorized Kerosene	80 fl oz
Methyl Salicylate	1 fl oz
Pyrethrum Powder	10 oz

Mix thoroughly by stirring from time to time; allow to stand covered overnight and then filter through muslin.

This spray is inflammable and should not be used near open flames.

Deodorant Spray

(For public buildings, sick rooms, lavatories, etc.)

Pine-Needle Oil	2
Formaldehyde	2
Acetone	6
*Isopropyl Alcohol	20

1 oz of this mixture is diluted with 1 pt. water for spraying.

Cresol Disinfectant

| †Caustic Soda | 25.5 g |
| Water | 140.0 cc |

Dissolve in a Pyrex or enameled dish and warm. To this, add slowly the following warmed mixture:

| **Cresylic Acid | 500.0 cc |
| Rosin | 170.0 g |

Stir until dissolved and add water to make 1,000 cc.

---

*Inflammable.
**Poison.
†Do not get this on the skin as it is corrosive.

Ant Poison

Sugar	1 lb
Water	1 qt
**Arsenate of Soda	125 g

Boil and stir until uniform; strain through muslin and add 1 spoonful honey.

Bedbug Exterminator

*Kerosene	90 fl oz
Clove Oil	5 fl oz
**Cresol	1 fl oz
Pine Oil	4 fl oz

Simply mix and bottle.

Nonstaining Mothproofing Fluid

Sodium Aluminum Silicofluoride	0.50
Water	98.00
Glycerin	0.50
"Sulfatate" (Wetting Agent)	0.25

Stir until dissolved.

Fly Paper

Rosin	32
Rosin Oil	20
Castor Oil	8

Heat this mixture in an aluminum or enameled pot on a gas stove, with stirring, until all the rosin has melted and dissolved. While hot, pour on firm paper sheets of suitable size which have been brushed with soap water just before coating. Smooth out the coating with a long knife or piece of thin flat wood and allow to cool. If a heavier coating is desired, increase

---

**Poison.

the amount of rosin. Similarly, a thinner coating results by reducing the amount of rosin. The finished paper should be laid flat and not exposed to undue heat.

Baking Powder

Bicarbonate of Soda	28
Monocalcium Phosphate	35
Corn Starch	27

Mix these powders thoroughly in a dry can by shaking and rolling for ½ hour. Pack into dry airtight tins as moisture will cause lumping.

Malted-Milk Powder

Powdered Malt Extract	5
Powdered Skim Milk	2
Powdered Sugar	3

Mix thoroughly by shaking and rolling in a dry can. Pack in an airtight container.

Cocoa-Malt Powder

Corn Sugar	55
Powdered Malt	19
Powdered Skim Milk	12½
Cocoa	13
Vanillin	⅛
Powdered Salt	⅜

Mix thoroughly and then run through a fine wire sieve.

Sweet Cocoa Powder

Cocoa	17½ oz
Powdered Sugar	32½ oz
Vanillin	¾ g

Mix thoroughly and sift.

Pure Lemon Extract

Lemon Oil U.S.P.	6½ fl oz
Alcohol	121½ fl oz

Shake together in 1-gal jug until dissolved.

Artificial Vanilla Flavor

Vanillin	¾ oz
Alcohol	2 pt

Stir the ingredients in a glass or china pitcher until dissolved. Then mix into the following solution:

Sugar	12 oz
Water	5¼ pt
Glycerin	1 pt

Color brown by adding sufficient burnt-sugar coloring.

Canary Food

Dried and Chopped Egg Yolk	2
Poppy Heads (Coarse Powder)	1
Cuttlefish Bone (Coarse Powder)	1
Powdered Soda Crackers	8

Mix well together.

Blue-Black Writing Ink

Naphthol Blue-Black	1 oz
Powdered Gum Arabic	½ oz
Carbolic Acid	¼ oz
Water	1 gal

Stir together in a glass or enameled vessel until dissolved.

Indelible Laundry-Marking Ink

A	Soda Ash	1 oz
	Powdered Gum Arabic	1 oz
	Water	10 fl oz

Stir until dissolved.

B	Silver Nitrate	4 oz
	Powdered Gum Arabic	4 oz
	Lampblack	2 oz
	Water	40 fl oz

Stir this in a glass or porcelain dish until dissolved. Do not expose the mixture to strong light or it will spoil. Then pour into a brown glass bottle. In using these solutions, wet the cloth with solution A and allow to dry. Then write on it with solution B using a quill pen.

Green Marking Crayon

Cresin	8
Carnauba Wax	7
Paraffin Wax	4
Beeswax	1
Talc	10
Chrome Green	3

Melt the first four ingredients in a container and then add the last two slowly, while stirring. Remove from the heat and continue stirring until thickening begins. Then pour into molds. If other-color crayons are desired, other pigments may be used. For example, for black, use carbon black or bone black; for blue, Prussian blue; for red, orange chrome yellow.

Antique Coloring for Copper

Copper Nitrate	4 oz
Acetic Acid	1 oz
Water	2 oz

Dissolve by stirring together in a glass or porcelain vessel. Pack into glass bottles.

Wet the copper to be colored and apply the coloring solution hot.

Blue-Black Finish on Steel

A Place the object in molten sodium nitrate at 700 to 800°F for 2 to 3 min. Remove and allow to cool somewhat, wash in hot water, dry, and oil with mineral or linseed oil.

B Then put the object in the following solution for 15 min:

Copper Sulfate	½ oz
Iron Chloride	1 lb
*Hydrochloric Acid	4 oz
**Nitric Acid	½ oz
Water	1 gal

Allow to dry for several hours. Place in a solution again for 15 min, remove and dry for 10 hr. Place in boiling water for ½ hr, dry, and scratch-brush very lightly. Oil with mineral or linseed oil and wipe dry.

Rust-Prevention Compound

Lanolin	1
*Naphtha	2

Mix until dissolved.

The metal to be protected is cleaned with a dry cloth and then coated with the composition.

Metal Polish

Naphtha	62 oz
Oleic Acid	⅓ oz
Abrasive	7 oz
Triethanolamine Oleate	⅓ oz
Ammonia (26%)	1 oz
Water	1 gal

---

*Inflammable
**Corrosive

In one container mix together the naphtha and oleic acid to a clear solution. Dissolve the triethanolamine oleate in the water separately, stir in the abrasive, and then add the naphtha solution. Stir the resulting mixture at a high speed until a uniform creamy emulsion results. Then add the ammonia and mix well, but do not agitate so vigorously as before.

Glass-Etching Fluid

Hot Water	12
†Ammonium Bifluoride	15
Oxalic Acid	8
Ammonium Sulfate	10
Glycerin	40
Barium Sulfate	15

Warm the washed glass slightly before writing on it with this fluid. Allow the fluid to act on the glass for about 2 min.

This is an excellent preservative for leather book bindings, luggage, and other leather goods.

White-Shoe Dressing

Lithopone	19 oz
Titanium Dioxide	1 oz
Bleached Shellac	3 oz
Ammonium Hydroxide	¼ fl oz
Water	25 fl oz
Glycerin	1 oz

Dissolve the last four ingredients by mixing in a porcelain vessel. When dissolved stir in the first two pigments. Keep in stoppered bottle and shake before using.

Waterproofing for Shoes

Wool Grease	8
Dark Petrolatum	4
Paraffin Wax	4

Melt together in any container.

---

†Corrosive.

Leather Preservative

Cold-Pressed Neatsfoot

Oil, Mineral	10
Castor Oil	10

Mix.

Polishes

Polishes are generally used to restore the original luster and finish of a smooth surface. They are also expected to clean the surface and to prevent corrosion or deterioration. There is no one polish which will give good results on all surfaces.

Most polishes contain oil or wax for their lustering or polishing properties. Oil polishes are easy to apply, but the surfaces on which they are used attract dust and show finger marks. Wax polishes are more difficult to apply, but are more lasting.

Oil or wax polishes are of two types: waterless and aqueous. The former are clear or translucent, the latter are milky in appearance.

For use on metals, abrasives of various kinds, such as tripoli, silica dust, or infusorial earth, are incorporated to grind away oxide films or corrosion products.

Black Shoe Polish

Carnauba Wax	5½ oz
Crude Montan Wax	5½ oz

Melt together in a double boiler. (The water in the outer container should be boiling.) Then stir in the following melted and dissolved mixture:

Stearic Acid	2 oz
Nigrosine Base	1 oz
Then stir in Ceresin	15 oz

Remove all flames and run in slowly, while stirring.

Turpentine	90 fl oz

Allow the mixture to cool to 105°F. and pour into airtight tins which should stand undisturbed overnight.

Clear Oil-Type Auto Polish

Paraffin Oil	5 pt
Raw Linseed Oil	2 pt
China-Wood Oil	½ pt
*Benzol	¼ pt
*Kerosene	¼ pt
Amyl Acetate	1 tbsp

Mix together in a glass jar and keep it stoppered.

Paste-Type Auto and Floor Wax

Yellow Beeswax	1 oz
Ceresin	2½ oz
Carnauba Wax	4½ oz
Montan Wax	1¼ oz
*Naphtha or Mineral Spirits	1 pt
*Turpentine	2 oz
Pine Oil	½ oz

Pour into cans or bottles which are closed tightly to prevent evaporation.

Floor Oil

Mineral Oil	46 fl oz
Beeswax	½ oz
Carnauba Wax	1 oz

Heat together in double boiler until dissolved (clear). Turn off the flame and stir in

*Turpentine	3 fl oz

---

*Inflammable—keep away from flames.

Lubricants

Lubricants, in the form of oils or greases, are used to prevent friction and wearing of parts which are rubbed together. Lubricants must be chosen to fit specific uses. They consist of oils and fats often compounded with soaps and other unctuous substances. For heavy duty, heavy oils or greases are used and light oils are suitable for light duty.

Gum Lubricant

White Petrolatum	15 oz
Acid-Free Bone Oil	5 oz

Warm gently and mix together.

Graphite Grease

Ceresin	7 oz
Tallow	7 oz

Warm together and gradually work in with a stick:

Graphite	3 oz

Stir until uniform and pack in tins when thickening begins.

Penetrating Oil
(For loosening rusted bolts, screws, etc.)

Kerosene	2 oz
Thin Mineral Oil	7 oz
Secondary Butyl Alcohol	1 oz

Mix and keep in a stoppered bottle.

Molding Compound

White Glue	13 lb
Rosin	13 lb

Raw Linseed Oil	⅓ qt
Glycerin	1 qt
Whiting	19 lb

Heat the white glue until it melts. Then cook separately the rosin and raw linseed oil until the rosin is dissolved. Add the rosin, oil, and glycerin to the glue, stirring in the whiting until the mass reaches the consistency of a putty. Keep the mixture hot.

Press this mass into the die firmly and allow it to cool slightly before removing. The finished product is ready to use within a few hours after removal. Suitable pigments may be added to secure brown, red, black, or other color.

In applying ornaments made of this composition to a wood surface, they are first steamed to make them flexible; in this condition, they will adhere to the wood easily and securely. They can be bent to any shape, and no nails are required for applying them.

Grafting Wax

Wool Grease	11
Rosin	22
Paraffin Wax	6
Beeswax	4
Japan Wax	1
Rosin Oil	9
Pine Oil	1

Melt together until clear and pour into tins. This composition can be made thinner by increasing the amount of rosin oil and thicker by decreasing it.

Candles

Paraffin Wax	30.0
Stearic Acid	17.5
Beeswax	2.5

Melt together and stir until clear. If colored candles are desired, add a very small amount of any oil-soluble dye. Pour into vertical molds in which wicks are hung.

Adhesives

Adhesives are sticky substances used to unite two surfaces. Adhesives are specifically called glues, pastes, cements, mucilages, lutes, etc. For different uses different types are required.

Wall-Patching Plaster

Plaster of Paris	32
Dextrin	4
Pumice Powder	4

Mix thoroughly by shaking and rolling in a dry container. Keep away from moisture.

Cement-Floor Hardener

Magnesium Fluosilicate	1 lb
Water	15 pt

Mix until dissolved.
The cement should first be washed with clean water and then drenched with this solution.

Paperhanger's Paste

White or Fish Glue	4 oz
Cold Water	8 oz
Venice Turpentine	2 fl oz
Rye Flour	1 lb
Cold Water	16 fl oz
Boiling Water	64 fl oz

Soak the glue in the first amount of cold water for 4 hr. Dissolve on a waterbath (glue-pot) and while hot stir in the Venice Turpentine. Use a cheap grade of rye or wheat flour, mix thoroughly with the second amount of cold water to about the consistency of dough or a little thinner, being careful to remove all lumps. Stir in 1 tbsp of powdered alum to 1 qt flour, then pour in the boiling water, stirring rapidly until the flour is thoroughly cooked. Let this cool and finally add the glue solution. This makes a very strong paste which will also adhere to a painted surface, owing to the Venice turpentine content.

Aquarium Cement

Litharge	10
Plaster of Paris	10
Powdered Rosin	1
Dry White Sand	10
Boiled Linseed Oil	Sufficient

Mix all together in the dry state, and make a stiff putty with the oil just before use.

Do not fill the aquarium for 3 days after cementing. This cement hardens under water, and will stick to wood, stone, metal, or glass and as it resists the action of sea water, it is useful for marine *aquaria*.

Wood-Dough Plastic

*Collodion	86
Powdered Ester Gum	9
Wood Flour	30

Allow the first two ingredients to stand until dissolved, stirring from time to time. Then, while stirring, add the wood flour, a little at a time, until uniform. This product can be made softer by adding more collodion.

Putty

Whiting	80
Raw Linseed Oil	16

Rub together until smooth. Keep in a closed container.

---

*Inflammable—keep away from flames.

Wood-Flour Bleach

Sodium Metasilacate	90
Sodium Perborate	10

Mix thoroughly and keep dry in a closed can. Use 1 lb to 1 gal boiling water. Mop or brush on the floor, allow to stand ½ hr, then rub off and rinse well with water.

*Paint Remover

Benzol	5 pt
Ethyl Acetate	3 pt
Butyl Acetate	2 pt
Paraffin Wax	½ lb

Stir together until dissolved.

Soaps and Cleaners

Soaps are made from a fat or fatty acid and an alkali. They lather and produce a foam which entraps dirt and grease. There are many kinds of soaps.

Cleaners contain a solvent, such as naphtha, with or without a soap. Abrasive cleaners are soap pastes containing powdered pumice, stone, silica, etc.

Concentrated Liquid Soap

Water	11
*Solid Caustic Potash	1
Glycerin	4
Red Oil (Oleic Acid)	4

Dissolve the caustic soda in water, add the glycerin, and bring to a boil in an enameled pot. Remove from the heat, add the red oil slowly, while stirring. If a more neutral soap is wanted, use more red oil.

---

*Do not get on the skin as it is corrosive.

Saddle Soap

Beeswax	5.0
*Caustic Potash	0.8
Water	8.0

Boil for 5 min, while stirring. In another vessel heat:

Castile Soap	1.6
Water	8.0

Mix the two solutions with good stirring; remove from the heat and add, while stirring:

Turpentine	12

Mechanics' Hand-Soap
Paste

Water	1.8 qt
White Soap Chips	1.5 lb
Glycerin	2.4 oz
Borax	2.4 oz
Borax	6.0 oz
Dry Sodium Carbonate	3.0 oz
Course Pumice Powder	2.2 lb
Safrol	To suit

Dissolve the soap in two-thirds of the water by heat. Dissolve the last three ingredients in the rest of the water. Pour the two solutions together and stir well. When it begins to thicken, sift in the pumice, stirring constantly till thick, then pour into cans. Vary the amount of water, for heavier or softer paste. Water cannot be added to the finished soap.

Dry-Cleaning Fluid

Glycol Oleate	2 fl oz
Carbon Tetrachloride	60 fl oz

---

*Do not get on the skin as it is corrosive.

Naphtha	20 fl oz
Benzene	18 fl oz

This is an excellent cleaner that will not injure the finest fabrics.

Wall-Paper Cleaner

Whiting	10 lb
Calcined Magnesia	2 lb
Fuller's Earth	2 lb
Powdered Pumice	12 oz
Lemenone or Citronella Oil	4 oz

Mix well together.

Household Cleaner

Soap Powder	2
Soda Ash	3
Trisodium Phosphate	40
Finely Ground Silica	55

Mix well and pack in the usual containers.

Window Cleanser

Castile Soap	2
Water	5
Chalk	4
French Chalk	3
Tripoli Powder	2
Petroleum Spirits	5

Mix well and pack in tight containers.

Straw-Hat Cleaner

Sponge the hat with a solution of:

Sodium Hyposulfite	10 oz
Glycerin	5 oz
Alcohol	10 oz
Water	75 oz

Lay the hat aside in a damp place for 24 hr and then apply a mixture of:

Citric Acid	2 oz
Alcohol	10 oz
Water	90 oz

Press with a moderately hot iron after stiffening with gum water, if necessary.

Grease, Oil, Paint, and Lacquer Spot Remover

Alcohol	1
Ethyl Acetate	2
Butyl Acetate	2
Toluol	2
Carbon Tetrachloride	3

Place the garment with the spot over a piece of clean paper or cloth and wet the spot with this fluid; rub with a clean cloth toward the center of the spot. Use a clean section of cloth for rubbing and clean paper or cloth for each application of the fluid. This cleaner is inflammable and should be kept away from flames. Cleaners of this type should be used out of doors or in well ventilated rooms as the fumes are toxic.

Paint-Brush Cleaner

A	Kerosene	2.00
	Oleic Acid	1.00
B	Strong Liquid Ammonia (28%)	0.25
	Denatured Alcohol	0.25

Stir slowly B into A until a smooth mixture results. To clean brushes, pour into a can and leave the brushes in it overnight. In the morning, wash out with warm water.

Rust and Ink Remover

Immerse the part of the fabric with the rust or ink spot alternately in solutions A and B, rinsing with water after each immersion.

A	Ammonium Sulfide Solution	1
	Water	19
B	*Oxalic Acid	1
	Water	19

Javelle Water
(Laundry Bleach)

Bleaching Powder	2 oz
Soda Ash	2 oz
Water	5 gal

Mix well until the reaction is completed. Allow to settle overnight and siphon off the clear liquid.

Liquid Laundry Blue

Prussian Blue	1
Distilled Water	32
*Oxalic Acid	¼

Dissolve by mixing in a crock or wooden tub.

Glassine Paper

Paper is coated with or dipped in the following solution and then hung up to dry.

---

*Poisonous

Copal Gum	10 oz
Alcohol	30 fl oz
Castor Oil	1 fl oz

Dissolve by letting stand overnight in a covered jar and stirring the next day.

Waterproofing Paper and Fiberboard

The following composion and method of application will make un-calendered paper, fiberboard, and similar porous material waterproof.

Parrafin (M.P. about 130°F.)	22.5
Trihydroxyethylamine Stearate	3.0
Water	74.5

The paraffin wax is melted and the stearate added to it. The water is then heated to nearly the boiling point and vigorously agitated with a suitable mechanical stirring device while the mixture of melted wax and emulsifier is being slowly added. This mixture is cooled while it is stirred. The paper on fiberboard is coated on the side which is to be in contact with water. This method works most effectively on paper-pulp molded containers and has the advantage of being much cheaper than dipping in melted paraffin as only about one tenth as much paraffin is needed. In addition, the outside of the container is not greasy and can be printed on after treatment which is not the case when treating with melted wax.

*Waterproofing Liquid

Paraffin Wax	$\frac{2}{5}$ oz
Gum Dammar	$1\frac{1}{5}$ oz
Pure Rubber	$\frac{1}{8}$ oz
Benzol	13 oz
Carbon Tetrachloride	To make 1 gal

Dissolve the rubber in the benzol, add the other ingredients, and allow to dissolve.

This liquid is suitable for wearing apparel and wood. It is applied by brushing on two or more coats, allowing each to dry before applying another coat. Apply outdoors as vapors are inflammable and toxic.

---

*Inflammable.

Waterproofing Heavy Canvas

Raw Linseed Oil	1 gal
Crude Beeswax	13 oz
White Lead	1 lb
Rosin	12 oz

Heat, while stirring, until all lumps are removed and apply warm to the upper side of the canvas, wetting it with a sponge on the underside before application.

Waterproofing Cement

China-Wood Oil	10 oz
Fatty Acids	10 oz
Paraffin Wax	10 oz
Kerosene	2½ gal

Stir until dissolved. Paint or spray on cement walls, which must be dry.

Oil-and Greaseproofing
Paper and Fiberboard

This solution, applied by brush, spray, or dipping, will leave a thin film which is impervious to oil and grease. Applied to paper or fiber containers, it will enable them to retain oils and greases.

Starch	6.6
Caustic Soda	0.1
Glycerin	2.0
Sugar	0.6
Water	90.5
Sodium Salicylate	0.2

This caustic soda is dissolved in the water. Then the starch is made into a thick paste by adding a portion of this solution. The paste is then added to the water. The resulting mixture is placed on a water bath and heated to about 85°C, until all the starch granules have broken. The temperature is maintained about ½ hr longer at 85°C. The other substances are then added and thoroughly mixed. Less water may be used if applied hot and then a thicker coating will result.

Fireproof Paper

Ammonium Sulfate	8.00
Boric Acid	3.00
Borax	1.75
Water	100.00

The ingredients are mixed together in a gallon jug by shaking until dissolved.

The paper to be treated is dipped into this solution in a pan, until uniformly saturated. It is then taken out and hung up to dry. Wrinkles can be prevented by drying between cloths in a press.

Fireproofing Canvas

Ammonium Phosphate	1 lb
Ammonium Chloride	2 lb
Water	½ gal

Impregnate with the solution; squeeze out the excess, and dry. Washing or exposure to rain will remove fireproofing salts.

Fireproofing Light Fabrics

Borax	10 oz
Boric Acid	8 oz
Water	½ gal

Impregnate, squeeze, and dry. Fabrics so impregnated must be treated again after washing or exposure to rain as the fireproofing salts wash out easily.

Dry Fire Extinguisher

Ammonium Sulfate	15
Sodium Bicarbonate	9
Ammonium Phosphate	1
Red Ochre	2
"Silex"	23

Use powdered substances only. Mix well and pass through a fine sieve. Pack in tight containers to prevent lumping.

Fire-Extinguishing Liquid

Carbon Tetrachloride	95
Solvent Naphtha	5

The naphtha minimizes the development of toxic fumes when extinguishing fires.

Fire Kindler

Rosin or Pitch	10
Sawdust	10 or more

Melt, mix, and cast in forms.

Solidified Gasoline

*Gasoline	½ gal
Fine-Shaved White Soap	12 oz
Water	1 pt
Ammonia	5 oz

Heat the water, add the soap, mix and, when cool, add the ammonia. Then slowly work in the gasoline to form a semisolid mass.

Boiler Compound

Soda Ash	87
Trisodium Phosphate	10
Starch	1
Tannic Acid	2

Use powders, mix well, and then pass through a fine sieve.

---

*Inflammable.

Noncorrosive Soldering Flux

Powdered Rosin	1
Denatured Alcohol	4

Soak overnight and mix well.

Photographic Solutions

Developing Solution

Stock Solution A

Pyro	4 oz
Pure Sodium Bisulfite	280 gr
Potassium Bromide	32 gr
Distilled Water	64 oz

Dissolve in a glass or enamel dish.

Stock Solution B

Pure Sodium Sulfite	7 oz
Pure Sodium Carbonate	5 oz
Distilled Water	64 oz

Dissolve separately in a glass or enamel dish.
Use the following proportions:

Stock Solution A	2
Stock Solution B	2
Distilled Water	16

At 65°F., this developer requires about 8 min.

Acid-Hardening Fixing Bath

A	Sodium Hyposulfite	32
	Distilled Water	8

Stir until dissolved and then add the following chemicals in the order given, stirring each until dissolves:

B	Warm Distilled Water	2½
	Pure Sodium Sulfite	½
	Pure Acetic Acid (18%)	1½
	Potassium Alum Powder	½

Add B to A and store in dark bottles away from light.

Chapter II

ADHESIVES

Caulking Compounds

(Knife Grade)

FORMULA NO. 1

"Polybutene" H-100	8.23
Blown Soya Oil Z_4	16.16
Tall Oil Fatty Acids	0.49
Calcium Carbonate	47.15
Raw Soya Oil (A_5 viscosity)	2.40
Blown Soya Oil (Z_4 viscosity)	5.00
Bodied Linseed (Z_4 viscosity)	0.80
"Polybutene" H-100	4.50
Soya Fatty Acid	0.30
Calcium Carbonate	27.80
Marble Dust	55.70
Talc	3.50

Vehicle ingredients are blended in a heavy-duty sigma blade mixer for 20 min. Filler components are added in the order shown and mixed thoroughly over a 40-min period. Total processing time is approximately 1 h.

Talc	23.75
Mineral spirits	3.93
Cobalt Drier (6%)	0.29

Mix as above.

No. 2
(Gum Grade)

"Polybutene" H-1900	8.30
Butyl Solution	7.69
Hydrous Silicate	3.14
Denatured Alcohol	0.16
Hydrogenated Rosin Ester	0.61
Calcium Carbonate	54.44
"Polybutene" L-14	10.70
Talc	7.25
Aluminum Paste	2.54
Mineral Spirits	5.17

In a sigma blade mixer use the following steps: H-1900 Polybutene and butyl solution are added and mixed thoroughly (10-15 min). Alcohol and the thixotropes are then added and mixed until swelling and homogeneity of the mix is obtained (15-20 min). $CaCO_3$ is added slowly; if mixing becomes labored, add portions of L-14 polybutene. Mix to a smooth consistency (15-20 min). Add talc and remainder of L-14; mix until homogeneous (15-20 min). Add mineral spirits and aluminum paste; mix until thoroughly dispersed (5-10 min).

Contact Adhesive

	Dry	Wet
Dow Latex 283 (45%)	70.0	155.6
Aliphatic Petroleum Tackifier emulsion (45%)	30.0	66.2
Polyacrylate Thickener	0.5	4.5

Hot Melt Adhesives

	FORMULA NO. 1	NO. 2	NO. 3	NO. 4	NO. 5
Paraffin Wax (154 F. AMP)	70	66.5	63.0	59.5	56.0
"Elvax" 250	30	28.5	27.0	25.5	24.0
"Nevex" 100 Resin	0	5.0	10.0	15.0	20.0

	NO. 6	NO. 7	NO. 8	NO. 9
		(Pressure Sensitive)		
"Solprene" 418	100	100	100	100
"Wingtack" 95	150	150	150	150
"Amoco" Resin 18-210	40	–	–	40
"Amoco" Resin 18-290	–	40	40	–
"Amoco" Polybutene L-14	10	–	80	–
"Amoco" Polybutene H-300	–	–	–	80
"Amoco" Polybutene H-1500	–	80	–	–
"Irganox" 1010	2	2	2	2

	NO. 10
"Solprene" 418	100
"Wingtack" 95	150
"Irganox" 1010	2
Polybutenes	0–125
"Amoco" Resin 18-210	40

	NO. 11
"Kraton" 1107	100
"Wingtack" 95	100
"Irganox" 1010	1.5
Polybutene	10–80
"Amoco" Resin 18-210	20–80

	NO. 12	NO. 13	NO. 14	NO. 15	NO. 16
"Kraton" 1107	100	100	100	100	100
"Wingtack" 95	100	100	100	100	100
"Amoco" Resin 18-210	–	–	–	–	50
"Amoco" Resin 18-290	40	40	20	40	–
"Amoco" Polybutene L-14	60	–	–	–	–

"Amoco" Polybutene H-100	–	10	–	–	–
"Amoco" Polybutene H-1500	–	–	40	40	20
"Irganox" 1010	1	1	1	1	–
Antioxidant 330, DLTDP	–	–	–	–	5

No. 17

	Dry	Wet
"Dow" Latex XD-30223 (42%)	70.00	166.67
Coumarone-Indene Resin Emulsion (50%)	10.00	20.00
Hydrocarbon Resin Emulsion (62.5%)	10.00	16.00
Butyl Benzyl Phthalate (100%)	10.00	10.00
Polyacrylate Thickener (11%)	.45	4.00

Laminating Adhesive

FORMULA No. 1

	Dry	Wet
Dow Latex 283 (45%)	70.0	155.6
Copolymer Tackifier Emulsion* (55%)	30.0	54.5
Polyacrylate Thickener	0.25	2.2

*Alkylated pentadiene/acetate resin.

No. 2

	Dry	Wet
"Dow" XD-8609.01	100.0	208.0
Magnesium Hydroxide	5.0	5.0
"Elvanol" 71-30	1.5	7.5

No. 3

(Ignition Deterrent)

	Dry	Wet
"Dow" XD-8260.04	100.00	200.00
Antimony Trioxide	5.00	5.00
Alumina Trihydrate	100.00	100.00
Magnesium Hydroxide	5.00	5.00

67.74% solids

Sealants

(Architectural)

FORMULA	NO. 1	NO. 2	NO. 3	NO. 4	NO. 5
"Bucar" 5214	100	100	–	–	–
"Polysar" XL-50	–	–	100	100	–
"Aid"-10	–	–	–	–	100
"Amoco" H-100 Polybutene	143	–	200	–	267
"Amoco" H-300 Polybutene	–	121	–	200	–
"Keltrol"	12.1	22	50	50	50
Stearic Acid	2.2	2.2	–	–	–
"Thixatrol" GST	–	–	6.6	6.6	10
Calcium Carbonate	480	480	653	653	582
Talc	242	244	417	417	433
Titanium Dioxide	12.1	11	67	67	67
Mineral Spirits	110	132	166	166	150
Cobalt Drier (6%)	0.22	0.33	0.5	0.5	0.5
Antioxidant	–	–	2.0	2.0	2.0
"Super Beckacite" 2000	–	–	5.0	5.0	5.0

15 min	A third of the polybutene and remainder of calcium carbonate is added.
25 min	Remaining polybutene and half the talc are added.
30 min	Remaining talc, TiO_2, and half the solvent are added and the mass mixed until homogeneous.
35 min	Mix cobalt drier and remaining mineral spirits and add the mixture incrementally over a 10 min period.
50 min	Mass should be homogeneous and ready for dumping.

Compression ram used 0–35 min.

Mixing procedure for "Polysar" XL-50 formulations

Start	Steam on to approximately 250 F. add thixotrope.
1 min	Add polybutene, antioxidant and phenolic resin.
5 min	Steam off; incrementally add fillers and pigment.
15 min	Cold water on, add "Polysar" XL-50, lower ram.
30 min	Add copolymer incrementally.
50 min	Mix for 5 min.
55 min	Dump.

Automotive Sealing Tape

FORMULA	NO. 1	NO. 2	NO. 3	NO. 4	NO. 5	NO. 6
"Bucar" 5214	100	100	100	100	100	100
"Amoco" Polybutene	100	100	100	100	160	160
Stearic Acid	2	2	–	–	–	–
"Super Beckacite" 2000	–	–	20	20	20	20
Carbon Black Beads	90	90	140	140	140	140

0 min   With cold water on, charge rubber, stearic acid, ½ carbon black and ½ polybutene to mixer; position ram and mix.

5 min   Add ¼ carbon black and ¼ polybutene.

10 min   Add ¼ carbon black and 1/8 polybutene.

20 min   Turn off cold water and steam heat to 250 F.

30 min   Steam off, cold water on...

35 min   Add remaining polybutene.

40 min   Dump.

No. 7

Polyisobutylene	32.52
Oleic Acid	0.97
"Amoco" Polybutene H-300	23.43
Asbestos	24.39
Asbestos	16.26
TiO$_2$	2.43

Use a double arm dispersion blade Baker Perkins mixer. The polyisobutylene, oleic acid and H-300 polybutene are added to the mixer and allowed to mix for 15-20 min. The asbestos and titanium dioxide are then added incrementally in the order shown. A mix time of 10 min is allowed between each addition.

After adding the titanium dioxide the whole mass is mixed for 10 min. Total mix time is approximately 1 h and 15 min.

No. 8

Ethylene Propylene Terpolymer	40.40
"Amoco" Polybutene H-300	40.40
"Amoco" Resin 18-210	8.08
Zinc Oxide	2.02
Stearic Acid	0.41
Tetraethylthiuram Disulfide	0.27
2-Mercaptobenzothiazole	0.13
Sulfur	0.21
Carbon Black (FEF)	8.08

The EPDM is banded on a tight mill at 82 C (180 F) to 93 C (200 F) for 10 min. H-300 polybutene/carbon black slurry is prepared and added in increments and milled 30 min. Zinc oxide and stearic acid are added and dispersed (5-10 min). Resin 18-210 is added and dispersed (5-10 min). Accelerators are added and dispersed (5 min). Sulfur is added and milled for 5 min. Stock is then extruded into ½" X ½" tape. Stock is cured in a forced draft oven for 24 h before testing.

No. 9

(Nondrying)

"Amoco" H-300 Polybutene	24.01
Amorphous Polypropylene Homopolymer	5.05
Butyl Rubber	1.72
Clay	16.37
Calcium Carbonate	44.05
Diatomaceous Silica	4.00
Cotton Fiber	4.80

In the laboratory, the sealant is compounded in a sigma blade mixer. Amorphous polypropylene is premixed with twice its weight of polybutene to facilitate complete dispersion. (This premixing may not be required in commercial scale equipment.) Additions to the mixer are made in the order of listing in the formula. The entire mass is then mixed for 1 h after the last addition.

No. 10

Charge to high-shear, low-speed mixer (double-blade sigma or planetary-type) and mix for several min:

"Duramite"	543.14
"Thixatrol" ST	47.73
"Ti-Pure" R-901	23.59

Charge, then mix for 45 min:

"Acryloid" RAS-75- (at 83% solids)	397.65
"Acryloid" CS-1- (at 83% solids)	170.86

Charge, then mix for 15 min the premix:

Cobalt Naphthenate (6%)	0.60
Zinc Naphthenate (8%)	2.98
"Silane" A-174	1.30
Xylene	20.52
"Exkin" No. 2	0.50

No. 11

	Dry	Wet
Dow Latex XD-8986.01 (53%)	75.0	141.5
Dow Latex 283 (45%)	25.0	56.6
Antifoamer	1.0	1.0
$CaCO_3$	195.0	195.0
TiO_2	5.0	5.0
Ethylene Glycol	3.0	3.0
Butyl Benzyl Phthalate	10.0	10.0
"Dalpad A" Coalescing Agent	10.0	10.0

No. 12

(Glazing)

Thermoplastic Elastomer	3.84
Tackifying Resin	7.69
"Amoco" Resin 18-290	9.61
"Amoco" Polybutene H-1500	24.98
Calcium Carbonate	42.27
Silica Extender	9.61
TiO_2	1.92
Antioxidant	0.04
Stabilizer	0.04

PVC Adhesive

	FORMULA NO. 1	NO. 2
"Geon" 102, EP	100	100
"Santicizer" 711	50	—
"Santicizer" 154	—	50
"Drapex" 10.4	3	3
"Mark" WS	2	2

Polyvinyl Acetate Adhesive

	FORMULA NO. 1	NO. 2
Polyvinyl Acetate	50	50
"Santicizer" 160	10	–
"Santicizer" 154	–	10
Solvents, Tackifiers, etc.	40	40

Flame-Retardant Plastisol Adhesives

	FORMULA NO. 1	NO. 2	NO. 3
"Geon" 121	100	100	100
"Santicizer" 148	–	55	110
"Santicizer" 711	110	55	–
"Drapex" 10.4	5	5	5
"Mark" 462	2	2	2
"Magcarb" L	20	20	20
"Thermogard" S	–	4	4

	NO. 4	NO. 5	NO. 6
"Geon" 128	100	100	100
"Santicizer" 148	–	80	80
"Santicizer" 711	80	–	–
"Magcarb" L	–	–	20
"Thermogard" S	–	–	4
"Drapex" 10.4	5	5	5
"Mark" 462	2	2	2

	NO. 7	NO. 8
"Geon" 121	100	100
"Santicizer" 154	70	–
"Santicizer" 711	–	70
Stabilizers	7	7
"Magcarb" L	15	15
"Thermogard" S	5	5

Gypsum Board Binder

Add 0.075–0.15% based on stucco weight.

Briquet and Pellet Binder

Add 0.75–1% "Norlig" (dry solid basis).

Animal and Poultry Feed Binder

Add 1–2% "Norlig" to feed and pellet.

Chapter III

COATINGS

Emulsion Coating

FORMULA No. 1

(Semigloss)

"Arolon" 580 Resin (42% NV)	140.0
Ethoxy Ethanol	50.0
Defoamer	4.0
6% Cobalt Drier—Water Dispersible	1.0
Rutile Titanium Dioxide	300.0
Calcium Carbonate	25.0
Nonionic Wetting Agent	6.0

Disperse on high speed mill

Acrylic Emulsion (46% NV)	440.0
Thickener Solution	50.0
Defoamer	1.0
Water	75.0

No. 2

(Gray Enamel)

"Arolon" 580	160
Ethoxy Ethanol	50
Igepal CO-800	6

Defoamer	2
6% Cobalt Drier–Water Dispersible	1
Lampblack Dispersion	20
Rutile Titanium Dioxide	150
Acrylic Emulsion (46% NV)	480
Propylene Glycol	25
Defoamer	2
"Mobilco" M	10
Thickener	40
Water	25

Electrical Conformal Coating

	FORMULA NO. 1	NO. 2
"Kalene" 800	100	100
"Whitex" Clay	35	35
"Mistron" Vapor Talc	60	60
Toluene	75	75
PbO_2 (VFC)	—	30
GMF (70%)	4.4	—
Mixed Viscosity, Brookfield, cps. @25 C	25,000	
Thixotropy Index	6.2	

Cure Properties: press cure, 1 h @66 C, plus a post cure of 3 days @66 C.

Butadiene-Styrene Paint

(Interior Blue)

	FORMULA NO. 1	NO. 2
Rutile Titania	233.5	230.0
"Celite" 281	51.5	50.7
Clay, ASP-400	58.0	57.2
Phthalocyanine Blue Toner	12.7	12.5
"Surfynol" TG	1.83	1.77
"Borden" PMX Special	0	—

NH₄OH (58%)	2.92	–
"Dowicide" A & G, 50/50	1.46	–
"Tamol" 731 (100% basis)	–	1.07
K₂CO₃	–	1.77
Latex, Firestone PL-11	428.0	422.5
3% Solution, "Methocel" 4000 cps	–	64.5
Water	284.00	240.0

Preparation of Casein Ammoniate

Casein	14.4
"Dowicide" A & G, 50/50	1.42
NH₄OH (58%)	0.72
Water	69.5

Preparation of Pigment Paste:

Disperse 1.83 parts of "Surfynol" TG surfactant in 168 parts of water with stirring. Stir in 33 parts of casein ammoniate solution and with continued stirring, add the pigments including the blue toner. This paste should have a consistency of 70-90 K.U. Transfer to a pebble mill and grind for 14 h. For other grinding equipment, adjust the viscosity as desired by adding either water or casein ammoniate.

Let-Down:

Transfer a weighed amount of the ground paste to a vessel equipped with a stirrer and add with stirring the remaining casein ammoniate, latex and water. Make certain each ingredient is completely mixed prior to making the next addition. Adjust the pH to 9 with 29% NH₄OH.

In Formula No. 2 use "Tamol" 731 instead of casein ammoniate and prepare the paint as shown previously for polyvinyl acetate.

Polyvinyl Acetate Paint

(Interior Blue)

	FORMULA NO. 1	NO. 2
Rutile Titania	236.0	220.0
"Celite" 281	52.1	48.5
ASP-400	58.9	54.9
Phthalocyanine Blue Toner	12.9	12.01

"Surfynol" TG	1.80	1.68
"Tamol" 731 (100% basis)	1.1	–
Lecithin	–	10.27
K_2CO_3	1.80	1.68
Ethyl "Carbitol"	23.8	22.2
"Flexbond" 800	446.0	
3% Solution, "Methocel" 4000 cps	66.5	109.0
Water	221.0	206.0

Preparation of Pigment Paste:

With stirring, add "Surfynol" TG surfactant, "Tamol" 731, K_2CO_3 and the pigments to 165 parts of water containing 33 parts of 3% "Methocel" solution. Grind in a pebble mill for 14 h. The viscosity should be 70–90 KU. For other grinding equipment, adjust the viscosity as desired by using more or less water.

When lecithin is used in place of "Tamol" 731, omit the "Methocel" in the grind and add the entire quantity required afterwards during let-down.

Let-Down:

Transfer a weighed amount of the ground paste to a vessel equipped with a stirrer. With stirring, add the polyvinyl acetate emulsion, Ethyl "Carbitol" and water to the paste. Adjust the viscosity by adding the "Methocel" solution or an equal amount of water. Adjust the pH to 7.5 or 8 by adding NH_4Cl if necessary.

Acrylic Paint

(Interior Blue)

	FORMULA NO. 1	NO. 2
Rutile Titania	233.0	230.0
"Celite" 281	51.25	50.5
"ASP"-400	58.0	57.2
Phthalocyanine Blue Toner	12.8	12.61
"Surfynol" TG	1.78	1.76
"Tamol" 731 (100% basis)	1.09	–

K_2CO_3	1.78	–
Ethyl "Carbitol"	23.2	22.77
Casein	–	14.20
NH_4OH (58%)	–	2.88
"Dowicide" A & G, 50/50	–	1.44
Acrylic Emulsion, Rhoplex AC-33	440.0	433.0
3% Solution, Methocel, 4000 cps	65.5	–
Water	218.0	276.5

Preparation of Pigment Paste:

With stirring, add "Tamol" 731, "Surfynol" TG surfactant, K_2CO_3 and the pigments to 184 parts of water containing 65.5 parts of 3% "Methocel" solution. Grind in a pebble mill for 14 h. The viscosity of the paste should be 70–90 KU. For other grinding methods, adjust the viscosity as desired by using more or less water.

Using a pebble mill, no color loss is noted if the "Methocel" is present during the grind. If part of the "Methocel" is added after grinding, some color loss is observed.

Let-Down:

Transfer a weighed amount of the ground paste to a vessel equipped with a stirrer and with stirring, gradually add the acrylic emulsion Ethyl "Carbitol" and water to the paste. Add NH_4OH if required to adjust the pH to 8.

The formulations as shown do not give foam when ordinary precautions are observed. If foam develops, "Surfynol" 104-E surfactant at .05 to .2% concentration based on the finished paint is very effective in elimination of foam. It may be stirred into the finished paint during let-down.

For Formula No. 2, the pigment paste preparation and the let-down procedure are similar to those for butadiene-styrene paints.

Metal Coatings

FORMULA NO. 1

"Carboset" XL-22 Resin (40% T.S.)	1500
Water	500

No. 2

"Carboset" XL-22 Resin (40% T.S.)	1500
Water	300
"Cymel" 301	52
PTSA Solution*	17.3
Ammonium Hydroxide (28%) added to pH 8.0 Resin	
Resin to crosslinker ratio	92/8

No. 3

"Carboset" XL-22 Resin (40% T.S.)	1500
Water	600
"Epon" 812	38.3
Triethylene Tetramine	1.9
Ammonium Hydroxide (28%) added to pH 8.0	
Resin to crosslinker ratio	94/6

No. 4

"Carboset" 514H Resin (40% T.S.)	1500
Water	500
Ammonium Hydroxide (28%) added to pH 7.5	

No. 5

"Carboset" 514H Resin (40% T.S.)	1575
Water	525
"Cymel" 301	111.2
PTSA Solution*	37
Ammonium Hydroxide (28%) added to pH 7.5	
Resin to crosslinker ratio	85/15

---

*p-toluene sulfonic acid (15%) in water –neutralize to 7.5 with NH₄OH

No. 6

"Petrolite" C-7500	50
"Petrolite" C-700	15
"A-C" 680	35
Oleic Acid	8
Zinc Octoate (19% zinc)	1.5
TEAE	12
Water	667

The above type formulation is prepared by the wax to water emulsification system.

Metal Primers

FORMULA No. 1

"VAGH" Vinyl Resin	64
MIBK	287
P-296-70 Soya Alkyd	158
Zinc Phosphate (65%)	230
R-TiO$_2$ (35%)	126
Xylene	144
Troy Chemical "ABC"	3
Cobalt Naphthenate (6%)	1.5
Lead Naphthenate (24%)	0.4
Calcium Naphthenate (5%)	0.3
"Troykyd" Antiskin	1.0

Dissolve "VAGH" in MIBK, add P-296-70 alkyd. When completely incorporated add pigment and pebble mill overnight to a 6H grind.

No. 2

Zinc Phosphate	300
Titanium Dioxide	160
Calcium Carbonate (30 micron)	120

"Bentone" 38	11
Methyl Alcohol	4
"P-296-60" Long Oil Soya Alkyd	243
Hi-Flash Naphtha	90
Raw Linseed Oil	80

Disperse in a Hi-Speed Mill.
Add

Mineral Spirits	120
Lead Naphthenate (24%)	2.50
Cobalt Naphthenate (6%)	1.05
Manganese Naphthenate (6%)	.50
ASA	1.00

No. 3

"Bakelite" VMCH	142.5
MIBK	213.3
Xylene	284.1
Cyclohexanone	70.8
R-900-TiO2	56.8
#317 Zinc Phosphate	49.8
"Flexol" 10-10	28.4

Dissolve VMCH in solvents. Charge into pebble mill, grind overnight.

Concrete Floor Coating

(To Prevent Dusting)

Zinc Sulfate	3
Water	8

Mop on floor and allow to dry.

Lacquer

FORMULA No. 1

(Brilliant White)

"Silikoftal" CC	62.6
Titanium Dioxide	37.2
Silicone Oil ATF	0.2

Dilute to 110-120 s efflux time DIN 4/20 C; as thinner a mixture of cyclohexanone, "Solvesso" 150 and isophoron (8 : 5 : 2) can, for example, be used.

No. 2

(Satin Finish)

"Silikoftal" CC	62.0
Titanium Dioxide (rutile)	34.5
"Syloid" 161	3.3
Silicone Oil ATF	0.2

Dilute to 110-120 s efflux time DIN 4/20 C.

Watch Dial Finish

(Dead White)

Sodium Chloride	16 g
Cream of Tartar	12 g
Silver Chloride	2 g
Silver Powder	1 g
Water	to make paste

Baking Slurry Coating

Polyphenylene Sulfide	100
Iron Oxide (brown)	16
TiO$_2$	16
Water	185
Propylene Glycol	60
Octyl Phenoxy Polyethoxy Ethanol	3

These ingredients are ball-milled 16 h, so the final coating passes 100% through a 200 mesh vibrating sieve (74 micron).

Tempera Paint

A	"Veegum" T	1.5
	Water	91.5
	Propylene Glycol	4.0
	Color	1.0
B	Propylene Glycol	1.0
	Titanium Dioxide	1.0
C	Preservative	q.s.

Add the "Veegum." T to the water slowly, agitating continually until smooth. Add propylene glycol and color with agitation. Mill B and add to A. Mix until uniform. Then add C.

Water Stains

	FORMULA NO. 1 (Sandalwood)	NO. 2 (Daffodil Gold)	NO. 3 (Misty Green)	NO. 4 (Desert Red)	NO. 5 (Monk Brown)	NO. 6 (Ox Blood)
Water	250.0	250.0	250.0	250.0	250.0	250.0
"Cellosize"	3.0	3.0	3.0	3.0	3.0	3.0
"Tamol" 850 Dispersant	4.5	4.5	4.5	4.5	4.5	4.5
"Igepal" CO-630 Surfactant	3.0	3.0	3.0	3.0	3.0	3.0
Potassium Tripolyphosphate	1.0	1.0	1.0	1.0	1.0	1.0
"Balab" 748 Antifoam	1.0	1.0	1.0	1.0	1.0	1.0
Ethylene Glycol	10.0	10.0	10.0	10.0	10.0	10.0
Preservative	2.0	2.0	2.0	2.0	2.0	2.0
Titanium Dioxide, Rutile	25.0	25.0	50.0	25.0	25.0	25.0
"Minex" 4 Extender	75.0	70.0	65.0	50.0	50.0	60.0
Zinc Oxide	50.0	65.0	50.0	50.0	50.0	60.0
"Celite" 281	25.0	25.0	25.0	25.0	25.0	25.0
Black Oxide	3.0	3.0	1.0	15.0	30.0	4.0
Brown Oxide	14.0	—	—	—	—	—
Yellow Oxide	25.0	35.0	8.0	—	—	—
Chromium Oxide	—	5.0	33.0	—	—	—
"Kroma" Red	—	—	—	65.0	—	25.0
Red Oxide	—	—	—	—	50.0	50.0
Burnt Umber	—	—	—	—	15.0	—
Butyl "Carbitol"	15.0	15.0	15.0	15.0	15.0	15.0
Water	180.0	196.0	210.0	195.0	213.0	220.0
"Ucar" Latex 366	320.0	305.0	335.0	305.0	335.0	300.0
"Balab" 748 Antifoam	2.0	2.0	2.0	2.0	2.0	2.0

Solvent Paint Remover

A	"Veegum" Pro	1
	"Klucel" M	1
B	Water	25
	N-Methyl-2-Pyrrolidone	73

Dry blend the "Veegum" Pro and "Klucel" and add to B slowly, agitating continually until smooth.

Directions for use: Apply evenly with brush to painted surface. Allow to stand for 10–30 min. Remove old finish with metal scraper or steel wool. Apply additional coats if necessary. Wipe wood surface with turpentine or denatured alcohol.

Paint Stripper for Metal

A	"Vegum" T	0.75
	"Kelzan"	0.25
	Water	62.00
B	"Ultrawet" 40 SX	1.00
	"Gafac" RE-610	2.50
C	Sodium Hydroxide	15.00
	Water	18.50

Dry blend the "Veegum" T and "Kelzan" and add to the water slowly, agitating continually until smooth. Add B to A. Combine C and add to A and B. Mix until uniform.

Directions for use: Apply liberally with a brush to painted metal surface. Allow to stand until old finish is loosened from surface (10–20 min). Remove old finish with scraper or steel wool. Rinse surface with water.

Caution: Contains caustic. Wear rubber gloves.

Evaporation Rate of Solvents and Thinners

Evaporation rates of solvents are not always in direct proportion to the boiling points. The following table illustrates the relative number of seconds required to evaporate to dryness an equal quantity of each solvent, at room temperature and atmospheric pressure at sea level.

Material	*Evaporation Time*
Acetone.	8 s
Methyl Acetone	9 s
Benzene.	15 s
Methanol	16 s
Ethyl Acetate	17 s
Methyl Ethyl Ketone	17 s
Heptane.	20 s
Lacquer Diluent Naphtha	21 s
Toluol.	42 s
Ethyl Alcohol.	43 s
Isopropyl Alcohol	43 s
VM&P Naphtha.	46 s
Methyl Isobutyl Ketone	60 s
Butyl Acetate	100 s
Xylol	143 s
Methyl Amyl Acetate.	213 s
Butanol.	222 s
Amyl Acetate	238 s
Hi Flash Naphtha	285 s
Cellosolve	312 s
Mineral Spirits	315 s
Diisobutyl Ketone	510 s
Diacetone Alcohol.	714 s
Butyl Cellosolve	1600 s
Isophorone.	3300 s

Chapter IV

COSMETICS

Depilatory

FORMULA No. 1

(Cream)

A	"Veegum"	3.5
	EDTA Na	1.0
	PEG 400	4.0
	Water	65.2
B	Mink Oil	4.0
	"Promulgen" D	8.8
	"Cetal"	1.0
C	Calcium Thioglycolate	5.5
	Calcium Hydroxide	6.5
D	Perfume	0.5

Heat A ingredients to 90 C and hold with continued high sheer agitation for 20 min. Then cool to 75 C and add premelted phase B. Gradually cool A and B with continued agitation down to 55 C. Add the two powders constituting phase C. Product will thin out as C is added. Use side-sweep agitation from here on gradually cooling further to 40-45 C before adding perfume. Take off at no higher than 35 C.

NO. 2

(Cream)

"Satulan"	1.5
"Crodacol" C	2.0
Mineral Oil	1.5
"Cosmowax" K	6.5
"Maprofix" WA	1.0
"Cellosize" QP 4400	0.2
Calcium Thioglycolate	8.0
Calcium Hydroxide	0.2
Water	79.1

Heat oil phase to 75 C. Heat water to 75 C, disperse "Cellosize" and stir until fully hydrated. Add water to oils with agitation and cool to 45 C. Add Calcium Thioglycolate and Hydroxide. Stir until uniform and fill into packages with minimum air space and good seal.

Note: As with all depilatories, contact with broken skin and with the eyes is to be avoided and use should be discontinued if burning sensations occur or if the individual has sensitive skin.

NO. 3

(Lotion)

A	"Veegum" HS	1
	Water	74
B	Propylene Glycol	5
C	"Amerchol" L-101	5
	"Solulan" 98	2
	"Arlacel" 165	2
D	Calcium Thioglycolate	5
	Calcium Hydroxide	6
	Perfume and preservative	q.s.

Add the "Veegum" HS to the water slowly, agitating continually until smooth. Add B and heat to 75 C. Heat C to 70 C. Add C to A and B with stirring. Cool to 40 C, add the calcium thioglycolate and stir; add the calcium hydroxide and stir until cool and uniform.

No. 4

(Gel)

"Pluronic" F-127 Polyol	20
Sodium Hydroxide	2
Water	78

No. 5

"Pluronic" F-127 Polyol	20
Calcium Thioglycolate (97%)	4
Sodium Hydroxide	1
Water	75

Place cold water (5–10 C) in container and add "Pluronic" F-127 polyol slowly with good agitation. Mix until all F-127 is dissolved, maintaining temperature at 10 C. Add depilatory ingredients, mix until homogeneous, and quickly transfer to containers. The product sets up into a ringing gel when it warms up to room temperature.

Hair Conditioner

FORMULA No. 1

(Aerosol)

"Lexamine" R-13		1.50
Citric Acid Monohydrate		0.50
"Lexgard" M		0.15
Perfume		0.05
Deionized Water, Dye	q.s. to	100.00

No. 2

"Lanexol" AWS	2.4
"Procetyl" 10	1.6
"Crotein" SPN or SPO	2.5
Ethanol SD40	46.5
"Klucel" HF	0.1
Distilled Water	46.9
Color, Preservative, Perfume	as needed

Dissolve "Klucel" in ethanol SD40 by heating and stirring. When dissolved, add "Lanexol" AWS, "Procetyl" 10, water, and "Crotein." For scalp odor suppression, 0.02–0.05% of antimicrobial may be incorporated in the "Lanexol"/"Procetyl." This should be dissolved by heating gently prior to addition of the other ingredients. Add color, preservative, and perfume.

No. 3

(Protein)

"Ammonyx" 4002	2.0
"Onamer" M	6.7
"Crotein" SPA	0.2
"Lantrol" AWS	0.5
SDA 40 Alcohol	10.0
Distilled Water	80.6

No. 4

(Cationic)

"Emulsynt" 610-A	8.0
"Onamer" M	6.7
Mineral Oil	2.0
Anhydrous Lanolin	2.0
White Petrolatum	q.s. to 100

No. 5

(Pump Type Spray)

"Gafquat" 734 Polymer	2.00
PVP/VA E-735 Copolymer	3.00
"Ammonyx" KP	0.30
"Tween" 20	0.15
Benzyl Alcohol	0.10
Ethanol SD40	66.00
Water	28.45

No. 6

(For Dry Hair – Balsam Type)

A	"Cerasynt" SD	3.50
	"Foamole" B	0.60
	"Ceraphyl" 424	1.00
	"Ceraphyl" 28	2.00
	Cetyl Alcohol	0.50
B	Deionized Water	86.55
	"Ceraphyl" 65	2.00
	"Ceraphyl" 60	1.00
	Lactic Acid 88% USP (10% aq. sol)	2.25
	"BTC" 2125M	0.10
	"Cellosize" QP (30,000)	0.50

Completely predisperse "Cellosize" in water, then add the rest of ingredients of B. Heat A and B to 80 C. Add A slowly to B with constant agitation at 80 C. Cool with stirring to 25-28 C.

Directions for use: First, shampoo hair, rinse well and leave wet. Then, (1) for dry damaged hair—apply approximately 1-2 tablespoons (amount depends on length of hair) of lotion directly to hair. Massage in one minute and rinse well. (2) for use as an all-purpose conditioning creme rinse—dilute 1 tablespoon (approx. ½ oz.) in a cup (8 oz.) of warm water and pass entire dilution throughout hair. Massage well and rinse.

No. 7

(Foaming)

"Lexaine" C		10.00
"Lexamine" C-13		6.00
"Lexamine" O-13		2.00
"Natrosol" 250 HHR		0.70
"Bronopol"		0.04
Perfume		0.15
Water, Dye q.s. to		100.00

No. 8

"Mirataine" CB	30.0
"Maprofix" ES	21.0
"Mirapol" A-15	2.1
Water	46.9

Adjust pH to 6.9.

No. 9

"Miranol$^{®}$" 2MCAS Modified	30.00
"Miranol$^{®}$" OS	15.00
"Mirapol" A-15	3.12
"Tween" 20	1.50
Water	50.38

Adjust pH to 6.6.

No. 10

"Miranol$^{®}$" 2MCAS Modified	25.00
"Miranol$^{®}$" OS	10.00
"Maprofix" ES	5.00
"Tween" 20	4.00
"Mirapol" A-15	2.10
PEG 6000 Distearate	0.75
Water	53.15

Adjust pH to 6.8.

No. 11

"Miranol$^{®}$" C2M Conc. NP	15.0
"Miranol$^{®}$" OS	15.0
"Maprofix" ES	15.0
"Mirapol" A-15	2.1

Super Amide L9C	2.0
"Tween" 20	1.5
Water	49.4

Adjust pH to 7.2–7.4.

No. 12

"Maprofix" ES	32.0
"Miranol®" C2M-SF Conc.	30.0
PEG 6000 Distearate	3.0
"Mirapol" A-15	1.6
Super Amide L9C	1.0
Water	32.4

Adjust pH to 6.4.

Damaged Hair Repair

FORMULA NO. 1

"Accobond" 3903	1
"Miramine" SH	1
Water q.s. ad	100

The above solution can be used to restore hair degraded and damaged by action of strong ammoniacal commercial bleach solutions.

NO. 2

"Accobond" 3903	5
"Hyamine" 3500 (cationic wetting agent)	2.5
Citric Acid	1
Water q.s. ad	500

No. 3

"Accobond" 3903	1
Dihydroxytrichloroethane	5
Polyethylene Glycol 600 Monolaurate	2
Glycerin	1
Water q.s. ad	120

This solution illustrates use of a composition with an alkylating setting solution. This combination of alkylating agent and methylol resin can find utility in nonoxidizing neutralizers for permanent waving.

No. 4

To Formula 3, add 60 cc's of 6% by weight aqueous solution of hydrogen peroxide. The resulting solution illustrates use of a composition with both an alkylating and oxidizing setting agent. The same results achieved with Example 3 are achievable with this.

A complete hair waving process is set forth below.

1. The hair is blocked off.
2. The hair is wetted with a mercaptan solution containing 6.6% by weight thioglycollic acid having a pH of 9.4.
3. The hair is wound up in curlers.
4. The hair is rewetted with the waving solution.
5. The hair is allowed to process for five to 15 min.
6. The hair is rinsed with lukewarm water.
7. A solution containing a composition in accordance with the second aspect of the present invention is applied to the hair.
8. The hair is allowed to stand for 5-10 min.
9. The hair is taken out of curlers.
10. The hair is rinsed.

Hair Relaxer

Ammonium Thioglycolate	8.00
"Carsonol" ALS	5.00
"Lanexol" AWS	2.00
Deionized Water	85.00

Blend under gentle heat.

Hair Thickener

"Gafquat®" 734 Polymer (GAF)	2.00
PVP/VA E0735 Copolymer	3.00
"Ceraphyl" 65 Emollient	0.20
Polypeptide AAS	3.00
"Solulan" 25	0.50
Ethanol SDA-40 (anhydrous)	30.00
Distilled Water	61.30
Perfume	q.s.

Antidandruff Gel

"Pluronic" F-127 Polyol	21
"Rewocid SBU-185"	4
Water	75
Perfume	q.s.

Dissolve antidandruff agent in cold water, add "Pluronic" F-127 polyol slowly with gentle mixing. Mix slowly until dissolved, keeping temperature below 10 C, or place in a refrigerator overnight. When clear solution forms, add perfume and transfer to containers. When solution warms to room temperature, a crystal clear, ringing gel will form.

Hair Setting Compounds

FORMULA NO. 1

"Lexamine" S-13	0.80
Lactic Acid (88%)	0.40
"Gafquat" 755	8.00
Cetyl Alcohol	0.50
Perfume	0.10
Water, Preservative, Dye q.s. to	100.00

Completely disperse the "Gafquat" in water. Add the remaining components, except preservative and perfume, and heat to 70 C. Stir until homogeneous. Cool and add preservative and perfume at 45 C.

	NO. 2 (Ultimate Hold)	NO. 3 (Super Hold)	NO. 4 (Regular Hold)
"Gantrez®" ES-225	7.0	5.50	2.50
2-Amino-2-Methyl Propanol	0.15	0.12	0.05
"Silicone" SF 1066	0.10	0.10	0.10
Perfume	0.10	0.10	0.10
SDA-40 Ethanol (190 proof)	92.65	94.18	97.25

For each formulation, first dissolve the neutralizing agent (AMP) in the alcohol. Add the resin. When the resin is thoroughly mixed, add the other ingredients in the order listed.

(Aerosol – Nonfluorocarbon)			
	NO. 5 (Ultimate Hold)	NO. 6 (Super Hold)	NO. 7 (Regular Hold)
"Gantrez" ES-225 or ES 425 Resins	5.50	5.00	4.00
AMP	0.11	0.10	0.08
"Ethoxylan" 100	0.10	0.10	0.10
Perfume	0.10	0.10	0.10
Solvent(s) *	74.19	74.70	75.72
Isobutane/Propane (90/10)	20.00	20.00	20.00

	NO. 8 (Super Hold)	NO. 9 (Regular Hold)
PVP/VA E735 or E635 or E535	5.00	4.00
Benzyl Alcohol	0.10	0.10
Silicone SF 1066	0.05	0.05

Perfume	0.10	0.10
Solvent(s)*	74.75	75.75
Isobutane/Propane (90/10)	20.00	20.00

No. 10

PVP K30	2.50
Benzyl Alcohol	0.10
Silicone SF 1066	0.05
Perfume	0.10
Solvent(s)*	77.25
Isobutane/Propane (90/10)	20.00

*Anhydrous ethanol, ethanol-water, ethanol-methylene chloride

	No. 11 (Ultimate Hold)	No. 12 (Super Hold)	No. 13 (Regular Hold)
Gantrez ES 225 or ES 425 (as is)	6.00	5.50	4.50
AMP	0.13	0.12	0.10
"Ethoxylan" 100	0.10	0.09	0.09
Perfume	0.10	0.10	0.10
Ethanol, SDA 40 (anhydrous)	93.67	94.19	95.21

	No. 14 (Super Hold)	No. 15 (Regular Hold)
PVP/VA E735 or 635 or 535	6.00	5.00
Benzyl Alcohol	0.12	0.10
Silicone SF 1066	0.06	0.05
Perfume	0.10	0.10
Ethanol SDA 40 (anhydrous)	93.72	94.75

No. 16

(Regular Hold)

PVP K30	3.00
Benzyl Alcohol	0.12
Silicone SF 1066	0.06
Perfume	0.10
Ethanol SDA 40 (anhydrous)	96.72

No. 17

(Pump Type – Protein)

"Onamer" M	6.7
"Procetyl" AWS	0.6
"Lanasan" Cl	0.2
SDA-40 Alcohol	47.8
Perfume	0.2
Distilled Water	44.5

No. 18

1.	"Polawax"	3.00
2.	Stearyl Alcohol	1.00
3.	Stearic Acid (triple-pressed)	1.50
4.	Cholesterol USP	0.60
5.	"Oat-Pro®"	2.00
6.	Lanolin (anhydrous)	2.00
7.	Mineral Oil (65/75 saybolt)	12.00
8.	PVP K-30	5.00
9.	Triethanolamine	0.40
10.	Methyl Paraben	0.10
11.	Propyl Paraben	0.05
12.	"Dowicil" 200	0.10
13.	Perfume	q.s.
14.	Deionized Water q.s.	100.00

Weigh and add 1, 2, 3, 4, 6, 7, and 11 into a container and begin heating and stirring. At the same time weigh and add 14 into a container and begin heating and stirring. Add 8 and stir until the material is completely dispersed, and the resultant dispersion is clear. Add the "Oat-Pro®" and continue stirring until complete dispersion occurs and no lumps can be felt or seen. Add 9 and 10 and bring the temperature to 70-73 C. Add the mineral oil-containing blend to the "Oat-Pro®"-water containing blend, when both are at 70-73 C. When all the mineral oil-containing blend has been added, begin cooling the batch. Cool to 40-43 C and add the remaining ingredients. Continue cooling to 30-35 C and package.

Hair Waves

FORMULA NO. 1

"Polawax"	5.0
"Polychol" 5	1.0
"Polychol" 10	4.0
Light Mineral Oil	20.0
Water	70.0

Add water at 75 C to the oil and emulsifiers, also at this same temperature, with rapid mechanical agitation. Continue stirring until cool and then homogenize for better stability. This concentrate may be added to the finished cold wave at a rate of 5% or even less. The water content of the cold wave is reduced to allow for the addition.

Concentrate, Above	5.0
Thiologycolic Acid	7.5
Ammonium Hydroxide (SG 0.88)	1.7
Water	to 100

NO. 2

"Volpo" 20	3.0
"Volpo" 10	4.0
"Crodacol" S	2.0
Ammonium Thioglycolate	10.0
Ammonium Hydroxide (28%)	7.0
Water	74.0

The "Volpo"s and stearyl alcohol, and the perfume are emulsified in about 50% of the water at 60 C. After cooling, the ammonia, the ammonium thioglycolate and balance of the water are added. The pH is adjusted to 9.5.

NO. 3

"Tetronic" 1508	20
Ammonium Thioglycolate (60%)	10
Water	70

Dissolve surfactant in cold water (5–10 C), mixing gently to avoid aeration. When solution is complete, add ammonium thioglycolate and mix until homogeneous. Transfer to suitable containers. Product sets up into a strong, ringing gel when warmed to room temperature.

NO. 4

"Pluronic" F-127	19.0
2-Mercaptoethylamine Hydrochloride	10.0
Water	71.0

Dissolve "Pluronic" F-127 in cold water (5–10 C), and add 2-mercaptoethylamine hydrochloride. Mix until a homogeneous solution is obtained, then transfer to containers. A strong, crystal clear, ringing gel forms on warming to room temperature.

NO. 5

"Manucol" KMOR	1.0
Alcohol (Industrial Methylated Spirits)	5.0
Glycerin	1.0
Preservative	as required
Color and Perfume	as required
Water	to 100

Disperse the "Manucol" KMR preservative in the alcohol and glycerin. Heat the water to 60 C (140 F) and then add the "Manucol" KMR dis-person with good stirring. Add the color and perfume on cooling.

NO. 6

"Gafquat" 734 (GAF)	3.50
"Lexamine" R-13	0.40
Citric Acid Monohydrate	0.10
Perfume	0.20
"Bronopol"	0.10
"Lexein" X250	1.00
Water, Dye q.s. to	100.00

Combine all ingredients except perfume, heat to 40 C with stirring until dissolved. Add perfume, to batch at 40 C. Stir and cool batch to 30 C, add water to make up evaporation loss, and fill at 30 C.

NO. 7

A	"Veegum" HS	1
	Water	69
B	Cetyl Alcohol	1
	Mineral Oil	5
	"Amerchol" L-101	5
	"Solulan" 98	2
	"Arlacel" 165	3
C	Ammonium Thioglycolate	12
	Ammonium Hydroxide	2
	Perfume and Preservative	q.s.

Add the "Veegum" HS to the water slowly, agitating continually until smooth. Heat to 80 C. Heat B to 75 C, add to A and mix until temperature reaches 35 C. AT 35 C, add C and mix slowly until uniform.

Hair Straightener

		FORMULA NO. 1	NO. 2	NO. 3
A	"Veegum" HS	2	2	2
	Water	52	53	57

B	Cetyl Alcohol	15	15	10
	White Petrolatum	5	5	5
	Mineral Oil (70 cs visc.)	8	8	5
	"Arlacel" 165	5	5	5
C	Ammonium Thioglyco-late (60%)	11	—	—
	Ammonium Hydroxide	2	—	—
	Sodium Hydroxide	—	2	—
	Sodium Bisulfite	—	—	2
	Water	—	10	10
	Ammonium Carbonate	—	—	4
	Perfume and Preservative	q.s.	q.s.	q.s.

Add the "Veegum" to the water slowly, agitating continually until smooth. Heat A to 80 C. Combine B and heat to 70 C. Add B to A and stir with rapid agitation until temperature cools to 35 C. Slowly dissolve ingredients in C and add A and B. Stir until cool and uniform.

Brilliantine

(Proteinized)

Light Mineral Oil	87.9
"Crotein" IPX	0.5
"Lanexol" AWS	2.0
"Procetyl" 10	9.6
Fragrance	q.s.

Hair Pomade

	FORMULA NO. 1	NO. 2
Paraffin Wax (140° m.p.)	5.0	14.9
"Crodamol" IPM	6.0	2.7
Mineral Oil	30.0	—
"Pentol"	—	57.4
Lanolin USP	10.0	2.2
Super Hartolan	3.0	—
Petrolatum, Yellow	46.0	—

White Petroleum Jelly	–	15.9
"Solan"	–	5.3
Propyl Paraben USP	0.1	0.1
Bergamot fragrance	–	0.2
Fragrance	q.s.	q.s.
Water	–	1.3

Heat all the ingredients except the fragrance until uniform. Cool, mix in the fragrance, package. A suitable dye for a green color is D & C Green No. 6 at 0.02%.

No. 3

1.	Microcrystalline Wax (165/175)		13.00
2.	Petrolatum (USP white)		100.00
3.	Mineral Oil (65/75 saybolt)	q.s.	33.35
4.	"Oat-Pro®"		2.00
5.	Propyl Paraben		0.10
6.	Perfume		q.s.

Weigh and add all the ingredients into a container, with the exception of 4 and 6. Begin heating while stirring. Heat this blend to 70–75 C, at which temperature sprinkle in the "Oat-Pro®." Continue stirring, and begin cooling the batch. Cool to 50–53 C, at which temperature the perfume should be added. Package at 48–52 C, while stirring continuously.

Hair Dye

FORMULA NO. 1

(Ash Blond)

A	"Emersol" 233LL	35.00
	"Tween" 81	10.00
	"Arlacel" 80	3.50
	"Atlas" G-1441	1.75
	"Centrolene" S	1.25

B	"Sequestrene" AA	0.10
	Sodium Sulfite	0.50
	Water	33.39
C	Ammonium Hydroxide (28%)	10.00
	Isopropanol	2.50
	p-Phenylenediamine	0.60
	o-Aminophenol	0.10
	p-Aminophenol	0.01
	4-Nitro-1, 2-Diaminobenzene	0.30
	Pyrogallol	0.70
	Resorcinol	0.20
	Hydroquinone	0.10
	Perfume	q.s.

Heat A to 70 C. Heat B to 72 C. Add B to A with stirring. Cool to room temperature with moderate stirring. Add the resulting unstable emulsion to dye solution (C) with gentle stirring until homogeneous.

No. 2

(Black)

A	"Lexemul" 515	7.00
	"Lexol" PG 8-10	4.00
	"Lexamine" S-13	0.40
	Stearyl Alcohol	3.00
	"Brij" 35	0.50
B	Water	68.30
	Propylene Glycol	5.00
	Acetic Acid	0.80
C	Sulfur, precipitated	1.00
D	Water	9.00
	Lead Acetate	1.00

Heat A and B to 80 C, disperse sulfur into B, add A slowly with agitation, cool until formula begins to get heavy and is smooth, add D very slowly and with very good agitation, cool with agitation to about 35 C.

Hair Bleach

FORMULA NO. 1

Cetyl Alcohol	2.5
"Kessco" GMS, SE (acid stable)	2.5
Hydrogen Peroxide (35%)	8.57 or 17.14
Deionized Water	q.s. to 100
Phosphoric Acid (10%)	as required
Fragrance	as needed

The two peroxide levels yield lotions with 3.0 and 6.0% active H_2O_2 respectively.

Adjust pH to 3.5–4.0 with phosphoric acid.

	NO. 2	NO. 3	NO 4
"Solan"	5.0	5.0	—
"Crodacol" C	5.0	—	—
"Crodacol" S	—	—	5.0
"Polawax"	—	5.0	—
"Polychol" 10	—	—	6.0
Hydrogen Peroxide	6.0	6.0	6.0
Water	84.0	84.0	83.0

Heat the oils to 50 C, the hydrogen peroxide is added to the water and warmed. The oils are added to the water with stirring. Note: The hydrogen perioxide content in these formulations was maintained at 3% and at 6% concentration level.

Antiperspirants, Roll-On

FORMULA NO. 1

"Solulan" 98	2.00
"Amerchol" L-101	5.00
Cetyl Alcohol	2.00
Glycerin	2.00
"Myrj" 52	4.00
"Veegum" HV	1.00
Water	48.00
"Chlorhydrol" 50%	36.00
Perfume and Preservative	q.s.

Disperse the "Veegum" in water with high speed mixing. Add the water phase at 70–80 C to the oil phase at 70–80 C while mixing. Continue mixing and cool to 40 C. Add the "Chlorhydrol."

No 2

A	"Veegum" HV	1.0
	Water	49.6
	"Methocel" E4M	0.4
B	SDA-40 Alcohol	8.0
	"Volatile Silicone" 7207	3.0
	"Arlamol" E	1.0
	"Brij" 97	1.0
C	Aluminum Chlorhydrate (50%)	36.0

No. 3

"Syncrowax" HGL-C	3.79
"Crodamol" PMP	71.96
"Procetyl" AWS	7.58
Water	1.52
"Micro-Dry"	15.15

Mix "Crodamol" PMP and "Syncrowax" HGL-C and heat to 75–80 C. Disperse "Micro-Dry" in above. Dissolve water in the "Procetyl" AWS, then add to oils. Perfume, cool and fill.

No. 4

A	"Lexemul" 561	13.00
	Cetyl Alcohol	0.50
	"Lexgard" P	0.05
B	Glycerin	3.00
	"Lexgard" M	0.15
	Water	43.30
C	Chlorhydrol (50%)	40.00
	Perfume	q.s.

Charge water, glycerin and "Lexgards" into a making tank and heat to 65-70 C with stirring until clear solution is formed. Melt "Lexemul" and cetyl alcohol together in a separate vessel, bring to 65-70 C and mix until homogeneous. Add B to A with vigorous agitation and cool with stirring to 50 C. Incorporate C, continue stirring and cool to 40 C. Cool to 30 C and package. The pH should be 7.0.

No. 5

A	"Nimlesterol" D	5.00
	"Emerest" 2408	4.00
	"Ethoxyol" AC	2.00
	USP Glycerin	2.00
	Cetyl Alcohol	1.00
B	"Veegum" HV	0.50
	Deionized Water	49.50
C	Basic Aluminum Bromide Complex	18.00
	Deionized Water	18.00
	Perfume and Preservative	q.s.

Disperse the "Veegum" HV thoroughly in water. Add the water phase B at 80 C to the oil phase A at 80 C, while stirring. Cool with stirring to 40 C. Add C to the combined AB. Continue stirring and cool to room temperature. (0.2% sorbic acid or "Dowicil" 200).

No. 6

A	Deionized Water	48.80
	Methyl *para* Hydroxybenzoate	0.12
B	"Arlamol" E	5.00
	Myristyl Alcohol	1.00
	"Brij" 72	3.00
	"Brij" 78 Polyoxyethylene 20 Stearyl Ether	2.00
	Propyl *para* Hydroxybenzoate	0.08
C	"Chlorhydrol"	40.00

Heat A to 75 C and B to 72 C. Add B to A with agitation. Stir to 45 C and add C. Continue stirring to room temperature.

	No. 7	No. 8	No. 9
"Polychol" 10	2.0	–	–
Super "Hartolan"	0.5	–	0.40
"Novol"	1.5	4.5	0.40
"Crodacol" C	–	1.0	0.75
"Volpo" S-2	–	3.0	–
"Volpo" S-20	–	3.5	–
"Volpo" 3	–	–	0.75
"Volpo" 10	–	–	0.75
"Polawax"	–	–	3.00
"Veegum" HV	0.5	0.5	0.50
Glycerin	2.0	2.0	2.00
"Arlacel" 165	4.0	–	–
Mineral Oil	4.0	–	3.00
Aluminum Chlorhydroxide (50%)	40.0	40.0	40.00
Perfume, Color	q.s.	q.s.	q.s.
Water	45.5	45.5	48.45

Disperse the "Veegum" in water at room temperature, heat to 90 C and keep at this temperature for approximately 1 h, stirring continuously. Replace any water lost by evaporation. Weigh out and melt the oils and water together, then add the "Veegum" solution to the oils at 80 C, with stirring. Cool to 45 C and add the Chlorhydrol solution, continue stirring and cool to room temperature. Add the fragrance and dyes.

No. 10

Powdered Aluminum Chlorohydrate	25.00
Talc	5.20
"Avicel" (pH 10.5)	69.55
Magnesium Stearate	.25
Fragrance	q.s.

Premix all powders and perfume with spinbar of PK blender for 10 min. Load mix into die and dry press at 1,100 psig; 15 s is the dwell time.

Antiperspirant Lotion

"Witconol" MST	2.0
Ceteareth-20	1.0
Polypropylene Glycol 400	7.0
"Emcol" E-607L	0.2
Aluminum Chlorohydroxide (50%)	40.0
Water	49.8
Perfume	q.s.

Dissolve "Emcol" E-607L in water, add aluminum chlorohydroxide solution and heat to 70–75 C. Add polyoxyethylene cetyl/stearyl ether and "Witconol" MST; stir until completely melted and uniformly dispersed. Maintain agitation while cooling to below 30 C. Add fragrance and package.

Cationic Quick-Breaking Foam Antiperspirant

A	"Chlorhydrol"	16.00
	Water	25.00
	Ethyl Alcohol	28.00
B	"Emcol" E-607S	2.25
	Cetyl Alcohol	0.75
	Ethyl Alcohol	28.00
	Concentrate	96.00
	Propellant A-70	4.00

Heat solutions separately to 60 C. Add solution B to solution A while stirring and continue stirring as solution cools. Pour into aerosol containers and pressurize.

Pumpable Antiperspirant Lotion

FORMULA NO. 1

A	Water	52.70
	Methyl *para* Hydroxybenzoate	0.12
	"Sequestrene®" NA$_2$	0.10

B	"Tween" 60 Polyoxyethylene 20 Sorbitan Monostearate	2.15
	"Arlacel" 60 Sorbitan Monostearate	0.85
	Isopropyl Myristate	4.00
	Propyl *para* Hydroxybenzoate	0.08
C	"Chlorhydrol®" (50%)	40.00

Part A: heat water to 70 C and dissolve the "Sequestrene" and the *para* hydroxybenzoate. Heat B to 72 C and add to A with mild agitation. Cool to 50 C and then add C slowly. Continue stirring and cool to room temperature. Package.

No. 2

"Rehydrol®"	20
Stearic Acid	2
Alcohol, SDA-40	76
"Arlamol" E	2

Dissolve "Rehydrol" and stearic acid in alcohol at room temperature. Add "Arlamol" E and stir until clear. Package.

Antiperspirant Cream

Formula No. 1

A	Mineral Oil #7	23.0
	"Pationic" CSL	3.2
	PEG 400 Dioleate	0.8
B	Glycerin	3.0
	Deionized Water	20.0
	Sodium Lactate (60%)	10.0
C	Aluminum Chlorohydrate (50%)	40.0
D	Perfume	q.s.

Heat A, B, and C to 70 C in separate vessels. Add B to C immediately before adding to A. Stir with moderate agitation to 40 C. Add D, finish mixing and package.

No. 2

Oil Phase:

"Amerchol" L-101	6.0 g
Beeswax	6.0 g
Paraffin Wax	11.7 g
Cetyl Alcohol	3.0 g
Mineral Oil	39.0 g

Aqueous Phase:

Methyl Parahydroxy Benzoate	0.1 g
Borax	1.2 g
Water	33.0 ml
Zinc or Aluminum Phenolsulfonate	3%-10%

Heat oil and aqueous phases to 70 C. Add aqueous phase to oil phase slowly with continuous stirring until room temperature. Mill in phenolsulfonate salt.

Antiperspirant Gel

"Tetronic" 1508	20.0
Aluminum Chlorohydrate (50%)	38.0
"Plurafac" RA-20 Surfactant	1.0
Perfume	0.4
Water	40.6

Dissolve "Tetronic" in a mixture of cold water (5-10 C) and aluminum chlorohydrate. Dissolve perfume in the "Plurafac" surfactant and add to water solution. Mix gently and transfer to suitable containers. Upon warming to room temperature, product sets up into a ringing gel.

Antiperspirant Aerosol

"Veegum" F	3.3
Cocoyl Sarcosine	1.0
"Chlorhydrol" Impalpable	4.0
Isopropyl Myristate	6.7
Propellant 12/11, 70/30	85.0

Mix components with agitation until uniform. Aerosol package.
Directions for use: Shake can before spraying. Hold about 10 in.
from underarm and spray.

Antiperspirant Stick

FORMULA NO. 1

A	"Avicel" PH 105	55.0
	Aluminum Chlorohydrate	23.0
	Talc (Italian)	15.0
	Zinc Stearate	1.75
B	Cyclomethicone	5.0
	Fragrance	0.25

Phase A powders are blended in a twin-cone blender for 5 min.

Phase B liquids are then atomized into the powder mix. Phase A/B placed in a Waring Blender at high speed for 2 min. Finally the entire blend is placed back in the twin-cone and mixing continues for 5 min.

Load powders into a compression die (a Carver Press for laboratory work) and press at 1700 psig.

For more uniform density and larger height to diameter ratio the powders may be loaded into a deformable envelope, sealed and pressed in an *isostatic chamber.* One such laboratory model is available for use at the FMC Princeton Forrestal Center.

NO. 2

"Novol"	3.00
Sodium Stearate	6.00
Propylene Glycol	5.00
Ethanol SD 40 (anhydrous)	86.00
Perfume, Color, and Germicide	q.s.

Weigh all ingredients except the perfume into a suitable jacketed mixer equipped with an efficient reflux condenser. Bring to about 75 C with agitation. When dissolved, begin cooling and add perfume at about 55 C. Pour into molds at 50 C. Explosion-proof equipment must be used throughout.

No. 3

"Witcamide" 70	28.5
"Witconol" APM	15.0
Propylene Glycol	30.0
"Rehydrol"	20.0
Water, Perfume	q.s.

Dissolve "Rehydrol" in water and propylene glycol at 25–30 C (77–82 F); use good agitation. Add "Witconol" APM and heat to 80 C (176 F). Add "Witcamide" 70 at 80–95 C (176–203 F) and stir until solution is clear. Cool to 77 C (171 F) with moderate agitation and add perfume. Package at 72–74 C (162–165 F).

Hardness or payout can be adjusted by raising or lowering the "Witcamide" 70 content.

No. 4

(Dry Compressed)

Ultrafine "Micro Dry"	25.00
Talc	9.00
"Avicel" Ph-105	65.75
Magnesium Stearate	.25

Deodorant, Personal

FORMULA No. 1

(Stick)

Sodium Stearoyl-2-Lactylate	10.0
Propylene Glycol	10.0
Water	39.0
SD 40 Alcohol	40.3
Sodium Capryl Lactylate	0.5
Perfume	0.2

pH adjusted to 9.5 with sodium hydroxide

No. 2

1.	"Chloracel" (40% w/w sol'n)		2.50
2.	Sodium Stearate		8.50
3.	Propylene Glycol		30.00
4.	Deionized Water	q.s.	100.00
5.	Ethyl Alcohol (SDA #40)		42.00
6.	Perfume		q.s.
7.	"Oat-Pro®"		2.00

Weigh and add 5 into a container and begin stirring using a variable speed agitator capable of imparting relatively high shearing stress. Heat 5 to 70-73 C and add 2. Weigh and add 3 to another container and begin stirring. Add 7 ("Oat-Pro®") and stir, by means of an agitator capable of imparting relatively high shearing stress until the "Oat-Pro®" is completely dispersed. Add the "Oat-Pro®" blend to the batch while maintaining the temperature at 70-73 C. Add the remaining ingredients with the exception of the perfume. Cool the batch to 60-63 C; continue stirring. Add perfume at 60 C and package.

No. 3

Zinc or Aluminum Phenolsulfonate	2.72
"Dowicil"	0.22
"Solulan" 16	0.91
Perfume	as needed
Ethyl Alcohol, SDA-40 (anhydrous)	96.15

No. 4

(Powder)

A	Sodium Bicarbonate	20.0
	Kaolin	30.0
	Zinc Stearate USP	4.0
	Zinc Ricinoleate	1.0
	"Avicel" PH-105 MCC	42.0
B	Perfume	0.5
	"Ultra Emulan"	2.5

Dry-mix A ingredients in ribbon blender or PK mixer. Then add B slowly to A during continuous agitation of A. Laboratory batches are conveniently made using Waring blender type equipment, while production-scale batches are ideally made in PK twin-cone blenders.

No. 5

(Powder)

1.	Talc	80.75
2.	Zinc Stearate	6.00
3.	"Syloid" 72	2.00
4.	"Oat-Pro®"	3.00
5.	Titanium Dioxide	2.00
6.	"Ottasept" Extra	0.25
7.	Perfume	q.s.
8.	Magnesium Carbonate	3.00

Add 2–8 to 1, and blend until the ingredients are uniformly dispersed.

No. 6

(Foot Spray)

Talc USP or Corn Starch	34.0
"Cab-O-Sil"	2.0
"Micro-Dry"	5.0
"Procetyl" AWS	4.0
PVP-VA	3.0
Menthol USP	0.5
Ethanol SD-40 (anhydrous)	51.5
Fill: Concentrate	15.0
Propellant 114	35.0
Propellant 12	50.0

Dissolve the Menthol, "Procetyl," and PVP in the alcohol and then disperse the Talc and "Micro-Dry." Add perfume, load into cans and gas.

Note: For antifungal purposes undecylenic acid or zinc undecylenate may be added to this system.

Bubble Bath

FORMULA NO. 1

		Order of Addition
"Lonzol" LA-300	20.0	(3)
Sodium Lauryl Ether Sulfate	15.0	(4)
"Amphoterge" SB	10.0	(5)
"Unamide®" LMDX	3.5	(6)
Citric Acid	0.5	(1)
Water	51.0	(2)

NO. 2

A	"Pationic" TEA-LMS	20.0
	"Clindrol" 100 LM	5.0
	"Pationic" ISL	3.0
	"Monateric" ISA-35	3.0
B	Methyl Paraben	0.2
	Deionized Water	68.3
C	Perfume Bouquet	0.5

Combine A, heat to 70 C. Combine B, heat to 72 C. Add B to A with agitation. Add C at 40 C. Stir to room temperature.

NO. 3

"Lakeway" 301-10	16
Hydroxyethyl Cellulose	1.0
Formalin	0.1
Phosphoric Acid	q.s. to pH 6.8-7.2
Water	Balance
Perfume and Dye	q.s.

No. 4

"Lakeway" 301-10	30
1 : 1 Coconut Diethanolamide	5
"Lakeway" 101-30	10
Sodium Chloride	2.0
Formalin	0.1
Phosphoric Acid	q.s. to pH 6.8-7.2
Water	Balance
Perfume and Dye	q.s.

Add water and salt first, then blend in other ingredients in order listed. (Note: for opacified version for a creme bubble bath, add 2% ethylene glycol monostearate.)

No. 5

Triethanolamine Lauryl Sulfate (40%)	5.00
Sodium Myristyl Ether Sulfate (60%)	15.00
Deionized Water	69.00
Sodium Chloride	4.00
"Lantrol" AWS	3.00
"Triton" X-102	2.00
Perfume	2.00
Formalin and Color	q.s.

Heat all ingredients in order to 75 C. Cool with stirring to 30 C.

No. 6

1.	Dupanol C		25.00
2.	Sodium Chloride	q.s.	100.00
3.	"Oat-Pro®"		10.00
4.	Perfume		q.s.
5.	CMC 7 LF		2.00
6.	"Dowacil" 200		0.10

Weigh and add 3 to a suitable blending apparatus. Begin stirring add 4. Mix until a uniform blend results. Add the remaining ingredients in order and continue blending. Mix until the ingredients are homogeneous in their distribution. Then package.

NO. 7

"Pluronic" L-121	1.0
Isopropyl Myristate	22.0
Isopropyl Palmitate	13.0
Light Mineral Oil	60.0
Perfume	4.0
Color	Trace

Place isopropyl myristate, isopropyl palmitate, "Pluronic" L-121 polyol, light mineral oil, perfume, and color in container. Gently mix a few minutes until product is clear and homogeneous. Transfer to containers. Final product is a clear fluid liquid. This floating bath oil spreads rapidly and evenly in warm water, from 45-60 C. The oil remains in a continuous film and thereby coats the body in an even manner. As the temperature drops, the oil exhibits much less tendency to break up into droplets, as compared to bath oils based on other types of spreading agents. The above formulation could be altered by varying the relative ratio of the isopropyl myristate, isopropyl palmitate mixture, and light mineral oil, or by replacing the light mineral oil with heavy mineral oil or another oil, suitably fortified with antioxidant, if needed.

After Bath Moisturizer

FORMULA NO. 1

Water	26.0
"Carbopol" 941	0.1
"Glucam" E-10	5.0
Specially Denatured Alcohol #40	60.4
"Glucamate" SSE-20	4.0
Perfume Oil	2.0
Diisopropyl Adipate	1.5
Diisopropanolamine (10% in water)	1.0
Color	q.s.

No. 2

A	Alcohol SD 40	70.0
	Sodium Lactate (60%)	5.0
	Isostearyl Lactate	3.0
	Perfume	2.7
	"Pationic" ISL	2.0
B	Deionized Water	17.3

Combine A and mix until clear. Slowly add B with agitation. Filter if desirable.

Bath Gelee

1.	"Sipon" LT 6		57.00
2.	"Emid"-6511		5.00
3.	"Amidox" C-2		5.00
4.	Propylene Glycol		5.00
5.	"Oat-Pro®"		2.00
6.	Methyl Paraben		0.10
7.	Propyl Paraben		0.10
8.	"Dowacil" 200		0.10
9.	Deionized Water	q.s.	100.00
10.	Perfume		q.s.

Weigh and add ingredient 1 into a container and begin stirring (Be careful to avoid aeration). Weigh and add 2, 3, 4, 6, and 7 and begin heating. Heat to 70–73 C while stirring carefully to avoid aeration. Weigh and add 9 to another container and begin stirring. A relatively high shear imparting stirrer should be used, a variable speed agitator with a propeller type stirrer is recommended. Begin heating and add 5. Stir until the "Oat-Pro®" is completely dispersed. Heat to 70–73 C. When both the "Oat-Pro®" containing blend and the Sipon LT-6 containing blend are both at 70–73 C, add the "Oat-Pro®" blend to the Sipon LT-6 blend. When all the "Oat-Pro®" blend has been added, begin cooling. Cool while continuing to stir to 40–43 C, at which temperature add the remaining ingredients. Cool to 25–30 C and package.

Body Scrub

A	"Chemical Base" 6532	0.50
	Lactic Acid (85%)	0.16
	Water	53.50
B	"Maprofix" TLS-500	30.00
	"Sandopan" TFL Conc.	10.40
	"Monamid" 150	3.00
	Sorbitol (70%)	2.00
	Perfume and Misc.	q.s.

Heat A to 80 C with stirring. Add B to A with stirring.

After Bath Powder

1.	Talc	q.s.	100.00
2.	Magnesium Carbonate		5.00
3.	Zinc Stearate		4.00
4.	Zinc Oxide		2.00
5.	"Oat-Pro®"		20.00
6.	Dioxin		0.10
7.	Perfume		q.s.

Weigh and add 1 into a container and begin stirring. Add 7 and stir until the perfume is completely dispersed. Add the remaining ingredients and stir until a uniform blend results. Package into suitable containers.

Talcum Powder

1.	Talc	70.00
2.	"Oat-Pro®"	15.00
3.	Zinc Stearate	9.00
4.	Perfume	q.s.
5.	Dioxin	0.10

Weigh and add 1 into the blending apparatus and begin stirring. Weigh in 4 and stir until a uniform blend results. Weigh and add 5 and the remaining ingredients while stirring continuously. Package into suitable containers.

Sun Screen

FORMULA NO. 1

A	"Veegum"	1.5
	Water	78.5
	Triethanolamine	4.0
B	"Myvacet" 9-40	4.0
	Cetyl Alcohol	0.5
	Stearic Acid	1.5
	Cocoyl Sarcosine	5.0
C	p-Aminobenzoic Acid	5.0
	Preservative	q.s.

Add the "Veegum" to the water slowly, agitating continually until smooth. Add the triethanolamine and heat to 65 C. Heat the components of B to 70 C. Add B to A with agitation. Add C and mix until cool.

This product should be packaged in an opaque container because exposure to light may cause discoloration.

NO. 2

"Amerscreen" P	1.0
"Ohlan"	0.5
"Solulan" PB-20	5.0
Isopropyl Palmitate	5.0
Stearic Acid XXX	6.5
"Arlacel" 165	6.0
Silicone Fluid 200 (350 cstks.)	1.0
"Veegum" HV	1.5
"Glucam" E-20	5.0
Water	68.5
Perfume and Preservative	q.s.

Disperse the "Veegum" in water with high speed mixing. Add the water phase at 85 C to the oil phase at 85 C while mixing. Continue mixing and cool to 30–35 C.

No. 3

"Poloxamine" 1508	20
Isopropyl Alcohol	18
Amyldimethyl PABA	1.5
Water	60.5
Preservative, Perfume	q.s.

No. 4

"Tetronic" 1508 Polyol	20.0
Isopropyl Alcohol	18.0
Escalol 506	1.5
Water	60.5
Preservative	q.s.
Perfume	q.s.

Place water, "Tetronic" polyol, alcohol, and "Escalol" 506 in vessel. Heat to 80 C and mix gently until homogeneous. Add perfume and preservative and transfer to suitable containers. Product sets up into a strong, clear gel when cooled to room temperature.

No. 5

A	"Arlamol" E	8.0
	"Arlasolve®" 200 Polyoxyethylene 20 Isohexadecyl	
	Ether	1.2
	"Brij®" 72 Polyoxyethylene 2 Stearyl Ether	2.8
	Stearyl Alcohol (USP)	4.0
	"Amerscreen" P	2.0
B	"Carbopol" 934	0.2
	"Dowicil" 200	0.1
	Deionized Water	81.3
C	Triethanolamine	0.2
D	Perfume	0.2

Disperse the "Carbopol" 934 in the water. Heat A to 60 C and B to 65 C. Add B to A with good agitation. Add C. Add D between 35–40 C. Pour about 35 C.

No. 6

A	1.	Mineral Oil		5.0
	2.	"Alcolec" PG		1.0
	3.	Cetyl Alcohol		1.1
	4.	Stearic Acid		0.7
	5.	"Neobee" M-20		1.5
	6.	"Arlacel" 60		0.5
	7.	"Tween" 60		1.5
	8.	"Escalol" 507		2.0
	9.	Preservative, oil soluble		q.s.
B	10.	Deionized Water		70.0
	11.	"Carbopol" 941		0.1
	12.	Glycerin		3.0
	13.	Triethanolamine		0.6
	14.	Preservative (water soluble)		q.s.
	15.	Deionized Water	q.s.	100.0
C	16.	Perfume		q.s.

Dissolve 2 in 1 and 5 at 60 C. Then add 3, 4, 6, 7, 8, and 9. Bring to 75 C. In a separate vessel, dissolve 11 in 10 at 75 C. When clear add 12-15. Add B to A at 75 C. Cool. Add C at 40-42 C.

No. 7

"Amerchol" L-101	5.0
"Amerlate" P	1.0
"Solulan" 25	3.0
"Carbopol" 934	.5
"Natrosol" 250HR	.2
Distilled Water	58.8
Triethanolamine	.5
"Amerscreen" P 80/20	2.5
Ethanol	28.5
Perfume and Preservative	as needed

1. Disperse the "Carbopol" in a portion of the water, using high-speed mixing.

2. Disperse the "Natrosol" separately in a portion of the water, using high-speed mixing. Heat to 80 C and hold until uniform and free from lumps.
3. Combine thickener suspensions from Steps 1 and 2 and heat to 75 C.
4. Add the water phase at 75 C to the oil phase at 75 C. Continue mixing for 5 min.
5. Add the triethanolamine. Continue mixing while cooling to 38 C.
6. Add the "Amerscreen" P dissolved in ethanol and cool to 30 C.

No. 8

(Aerosol)

"Polawax"	3.0
Ethanol SD-40 (95%)	54.6
"Escalol" 506	2.0
Water	30.4
Propellant 12/114 (40 : 60)	10.0

Cocoa Butter Tanning Lotion

A	"Veegum"	2.0
	Water	69.5
B	"Giv-Tan" F	1.5
	Sorbitan Monostearate	1.5
	Polysorbate 60	8.5
	Silicone 556 Fluid	2.0
	Cobee 76	3.0
	Cocoa Butter	12.0
	Preservative	q.s.

Add the "Veegum" to the water slowly, agitating continually until smooth. Heat to 75 C. Heat B to 70 C. Add A to B with agitation. Allow the lotion to cool to room temperature with stirring.

Water Resistant Suntan Products

		(Lotion)	*(Cream)*
A	"Veegum"	2.0	1.75
	Water	70.5	77.00
	Glycerin	3.5	2.25
B	"Giv-Tan" F	1.0	1.00
	Sorbitan Monolaurate	3.5	3.50
	Polysorbate 20	4.5	4.50
	Silicone 556 Fluid	5.0	5.00
	Stearic Acid	5.0	5.00
	Glyceryl Monostearate	5.0	—
	Perfume and Preservative	q.s.	q.s.

Add the "Veegum" to the water slowly, agitating continually until smooth. Add glycerin and heat to 70 C. Heat B to 75 C. Add B to A, then mix until cool.

Indoor Tanning Lotion

A	"Veegum"	4.0
	Sodium Carboxymethylcellulose (low visc.)	0.5
	Water	77.5
	Propylene Glycol	2.0
	Acetol	2.0
	Buffer Solution	q.s.
B	Dihydroxyacetone	4.0
C	Synthetic Iron Oxide	3.0
	Titanium Dioxide	3.0
D	Ethanol	4.0
	Preservative	q.s.

Dry blend the "Veegum" and CMC and add to the water slowly, agitating continually until smooth. Add the balance of A and mix. Buffer to pH 5.0 with saturated citric acid. Add B to A gradually with mixing. Blend C and then add A and B, mixing until uniform. Add D and mix until uniform.

After Sun Moisturizing

FORMULA NO. 1

A	1.	"Amerlate" W		2.00
	2.	Stearic Acid (triple pressed)		2.20
	3.	Glyceryl Monostearate (non self emulsifying)		1.50
	4.	Mineral Oil (65/75 saybolt)		12.00
	5.	Isopropyl Palmitate		3.00
	6.	BHA		0.10
	7.	Propyl Paraben		0.10
B	8.	Deionized Water	q.s.	100.00
	9.	"Oat-Pro®"		2.00
	10.	Propylene Glycol		5.00
	11.	Methyl Paraben		0.10
	12.	Triethanolamine		0.60
	13.	"Dowacil" 200		0.10
	14.	Perfume		q.s.

Weigh and add Part "A" to a container and begin stirring and heating. Heat to 70–73 C. Weigh and add 8 into another container and begin stirring and heating. A stirrer capable of imparting relatively high shearing stress is recommended. Add 9 ("Oat-Pro®") and stir until the "Oat-Pro®" is thoroughly dispersed (no lumps of "Oat-Pro®" should be evident). After the "Oat-Pro®" has been thoroughly dispersed add 10, 11, and 12. Bring the temperature of this blend, Part "B", to 70–73 C, and add Part "B" to Part "A", which should also be at 70–73 C. Start cooling the batch, while stirring continuously. Cool to 40–43 C and add the remaining ingredients. Continue cooling and stirring. Package at 25 C.

No. 2

Sesame Oil USP		15.0
"Polysynlane"		20.0
Glyceryl Monostearate		3.0
Isopropyl Myristate		10.0
"Carbopol" 934		0.2
Propylene Glycol		10.0
Triethanolamine		1.0
Anhydrous Lanolin		5.0
Perfume and Preservative		.8
Water	q.s.	100.00

No. 3

A	"Dipsal"	6.0
	"Schercomid" AME-70	5.0
	PEG-400 Monostearate	2.5
B	Micro Crystalline Cellulose	1.7
	Hydroxypropyl Cellulose	0.1
	"Polyox" WRSN-750	0.1
	Water	84.1
C	Perfume	0.5

Prepare B by sifting in all other ingredients into water at 70 C using extra-high sheer agitation for 10-15 min. Then add B to A at 70 C with agitating. Cool and stir to 45 C and add perfume.

Shampoo

FORMULA NO. 1

		Order of Addition
"Lonzol" LS-300	35.00	(4)
"Unamide®" CDX	4.00	(5)
Sodium Chloride	0.80	(3)
Citric Acid	0.17	(2)
Water	60.03	(1)

No. 2

(Low pH)

		Order of Addition
"Lonzol" LA-300	30.00	(4)
"Unamide®" CDX	3.00	(5)
Sodium Chloride	0.50	(3)
Citric Acid	0.25	(2)
Water	66.25	(1)

No. 3

(Pearly)

Sodium Lauryl Sulfate (30%)	30.00
"Lexaine" C	6.00
"Lexemul" EGMS	1.00
Sodium Chloride	1.40
Perfume	0.20
Water, Dye, Preservative q.s. to	100.00

Add sodium laurylsulfate, "Lexaine" C, "Lexemul" EGMS, and sodium chloride to water and heat to 65-70 C. When the EGMS is completely dispersed, cool to 45 C. Adjust pH to 6.0-6.5 with phosphoric acid, and add dye, preservative and perfume. Fill at 35 C.

No. 4

("Prell" Type)

Sodium Lauryl Sulfate (29%)	30.0
"Mazamide" LS-196	7.0
"Mapeg" 6000 DS	3.0
Methyl Paraben	.1
Deionized Water	60.0
Dye and Perfume to Suit	—

No. 5

(Mild)

A	Sodium Lauryl Sulfate (28–30%)	15.0
	"Pationic" TEA-LMS	4.5
	"Pationic" 138C	3.0
	Lauric Diethanolamide	3.0
	PEG 6000 Distearate	1.0
	Methyl Paraben	0.2
B	Deionized Water	73.3
	Fragrance	q.s.

Blend A and heat to 70 C. Heat water to 72 C. Add water to A with agitation. Continue to agitate while cooling to 40 C. Add fragrance and package.

No. 6

"Neodol" 25-3S (60%)	26.7
"Ammonyx" LO (30%)	13.3
Total Actives (100%)	20.0
Sodium Chloride	4.0
FD and C Yellow No. 5 (0.1 % aqueous sol'n)	1.5
Fragrance	0.2
Water	q.s.

No. 7

(High Viscosity)

"Lonza" SC-503Z	25.00
Sodium Chloride (salt)	0.20
"Dowicil" 200	0.05
Water, Dye, Perfume	74.75

Final viscosity = approx. 4000 cps

Dissolve the "Dowicil" 200 and sodium chloride in the water. Add "Lonza" SC-503Z with moderate agitation. When a clear viscous (approximately 4000 cps) mixture has been achieved, add perfume and dye.

No. 8

(Concentrated)

"Super Amide" GR	45.0
"Maprofix" TLS-500	45.0
Propylene Glycol	10.0

No. 9

(Concentrated)

"Anionyx" 12S	43.0
"Maprofix" WAQ	29.0
"Onyxol" 345	17.0
Propylene Glycol	3.0
Distilled Water	8.0

No. 10

(Concentrated)

"Duponol" QC	11.00
"Tegobetaine" C	4.00
"Arquad" C-50	0.75
"Klucel" MF	1.35
Methyl Parasept	0.10
Distilled Water	82.80

1. Weigh 188.1 g of an aqueous stock solution of "Klucel" MF (3.30% total solids) into a 1-liter beaker.
2. Work in (almost dropwise at the start) 29.7 ml of distilled water to Step 1, using gentle but thorough hand-stirring.
3. Place mixture on a finger pump equipped with ¼-in.-ID "Tygon" tubing, and let mix until uniform in appearance.
4. Add incrementally a total of 165.45 g of a clear "Duponol" QC/ methyl parasept stock solution and mix until the sample again appears uniform. This stock solution contains 1.05 g methyl parasept plus 385.0 g of as-received "Duponol" QC.
5. Lastly, add incrementally a total of 66.75 g of a clear "Tegobetaine"

C/"Arquad" C-50 stock solution and again mix until the sample preparation looks uniform in appearance. The sample becomes quite viscous after about one-third of the amphoteric/cationic surfactant stock has been added, and the pump has to be operated at near-maximum speed. The stock solution contains 140.0 g of "Tegobetaine" C plus 15.75 g of "Arquad" C-50.

No. 11

"Lakeway" 301-10	30
1 : 1 Coconut Diethanolamide	5
Propylene Glycol	0.5
Sodium Chloride	4-5
Phosphoric Acid	q.s. to pH 6.8-7.2
Formalin	0.1
Water	Balance
Perfume and Dye	q.s.

Add water first, then other ingredients in order listed.

No. 12

(Protein)

Triethanolamine Lauryl Sulfate	20.0
"Polymer" JR-30M	1.5
"Polymer" JR-400	0.1
Lauric Diethanolamide (high purity)	2.0
"Protein WSP" X-250	2.0
Preservative	0.1
Deionized Water, Perfume, Dye	74.3

Blend the powdered resin and disperse into a cold solution of the triethanolamine lauryl sulfate and the available water. Heat the dispersion gradually with stirring to 70–75 C, and then add the molten lauric diethanolamine. Maintain the temperature and stirring until the shampoo is homogeneous. When the shampoo has cooled to 35–40 C, stir the protein and preservative into the system. Finally, add the desired perfume and color.

NO. 13

(Protein)

Ammonium Lauryl Sulfate (30%)	25.00
"Lexaine" C	14.00
"Lexein" X250	5.00
"Bronopol"	0.05
"Lexgard" M	0.15
"Lexgard" P	0.05
Perfume	0.35
Sodium Chloride	0.35
Water, Dye q.s. to	q.s.

Add all components, except dye and perfume, to water. Heat to 45 C and stir until clear. Add perfume and dye. Adjust viscosity with sodium chloride, cool and fill.

		NO. 14	NO. 15
		(Antidandruff)	
A	"Veegum"	1.0	1.0
	Water	38.9	61.5
	Citric Acid	0.3	q.s.
	"Igepon" AC-78 (83% solids)	20.0	20.0
	"Titron" X-200 (28% solids)	27.8	—
	"Plurafac" C-17	—	5.0
B	"Modulan"	1.0	1.0
	Cetyl Alcohol	3.0	2.0
	Stearic Acid	6.0	—
	Glyceryl Monostearate A.S.	—	7.5
C	Vancide 89RE	2.0	2.0

Add the "Veegum" to the water slowly, agitating continually until smooth. Add the rest of A and heat to 85 C. Heat B to 90 C. Add B to A, mixing until cool. Add C to a small portion of the mixture and disperse thoroughly. Add this concentrate to the remainder of the mixture. Mix until uniform.

	No. 16	No. 17	No. 18
		(Antidandruff)	
Bentonite 670	3.00	3.00	3.00
"Super Amide" L-9	3.00	3.00	3.00
"Grocor" 5220	1.00	1.00	1.00
Titanium Dioxide	0.50	0.50	0.50
"Maprolyte" LX	55.00	—	—
"Maprofix" RH	—	—	55.00
"Maprolyte" 101	—	55.00	—
Selenium Sulfide	1.00	1.00	1.00
Citric Acid Powder	0.60	0.75	1.86
Sodium Acid Phosphate	0.30	0.30	0.30
Distilled Water q.s.	35.60	35.45	34.34

No. 19

(Coal Tar)

Stearic Acid T.P.	3.50
NaOH (50% sol'n)	1.40
"Onyxol" 42	6.00
"Miranol" C2M Conc.	10.00
"Super Amide" GR	1.00
Coal Tar Extract-alcoholic	5.00
"Maprofix" WA	50.00
Distilled Water	23.10

No. 20

(Sulfur)

"Veegum" HV (5% aqueous)	40.00
Sulfur (colloidal)	1.00
"Maprosyl" C	4.00
"Maprofix" TAS	40.00
Distilled Water	15.00

Hair Rinses

FORMULA NO. 1

Distilled Water	84.5
"Standapol" AB-45	5.0
"Dehyquart" A	5.0
Protein	5.0
"Natrosol" 250HHR	0.5
Perfume, Preservative, Dye, etc.	as needed

No. 2

(Protein)

A	"Solulan 25"	1.00
	Liquid Lanolin	0.50
	"Emcol" E-607S	1.50
	"Cerasynt" 1000D	2.00
	Cetyl Alcohol	0.50
B	Water	91.30
	"Maypon" 4CT	1.00
	Polypeptide AAS (20%)	2.00
C	Fragrance	0.20
	D & C yellow #10 (0.5%)	3 drops

Add B to A (both at 70 C) while stirring. Continue stirring; cool to 40 C and add fragrance.

Product increases in viscosity over 48 h. Agitation after this time will prevent future viscosity changes.

Use ½ oz to one cup of water.

No. 3

(Pearly)

"Emcol" E-607S	2.0
Cetyl Alcohol	1.0
Sodium Chloride	0.5
Water	96.5

Add "Emcol" E-607S to 30% of water and heat to 60 C. Melt cetyl alcohol and add to aqueous dispersion of "Emcol" E-607S at elevated temperature.

Cool, add balance of water containing salt. Add perfume and preservative as desired.

Mix one tablespoon in ½ cup of water and comb through clean, wet hair.

No. 4

(Creme)

"Arquad" 2HT-75	7.50
Glyceryl Monostearate	2.00
"Gafquat®" 755 Polymer	2.00
Distilled Water	87.99
Citric Acid (10% aqueous sol'n pH 5.0-5.5)	0.11
Glutaraldehyde (25%)	0.40

Add water to mixing tank. Add "Gafquat" polymer and mix until uniform. Add "Arquad" softener and glyceryl monostearate and heat mixture to 60 C with agitation. When batch is uniform, commence cooling. At 40 C add glutaraldehyde and perfume. Sample batch and calculate amount of citric acid needed to bring pH to 5.0-5.5.

No. 5

1.	Deionized Water	q.s.	100.00
2.	"Oat-Pro®"		2.00
3.	"Ammonyx"-4		7.00
4.	Cetyl Alcohol		1.50
5.	Sodium Chloride		0.80
6.	Perfume		q.s.
7.	Color		q.s.
8.	"Uvinul" MS-40		0.10
9.	"Dowicil" 200		0.10

Weigh and add 1 into a container and begin stirring, with a relatively high shear imparting propeller type stirrer. Add 2, which is the "Oat-Pro®", and stir until complete dispersion of the material takes place. Begin heating the water–"Oat-Pro®" blend, while stirring continuously. Heat this blend to 70-73 C. In another container weigh and add 3 and 4 and begin stirring and heating. Heat this blend to 70-73 C. When both blends are at 70-73 C add the "Ammonyx"-4–cetyl alcohol blend to the deionized water–"Oat-Pro®." When all of the "Ammonyx"-4–cetyl alcohol blend has been added, begin cooling the batch. Add 8 at 65 C. Continue cooling, and reduce the temperature to 40-43 C, at which time add the remaining ingredients. Stir until uniform and package.

Hair Spray

FORMULA NO. 1

(Pump Type)

"Onamer®" M	6.7
"Procetyl" AWS	0.6
"Lanasan" CL	0.2
SDA-40 Alcohol	47.8
Perfume	0.2
Distilled Water	44.5

Hair Setting Gel

1.	"Polawax"		3.00
2.	Stearyl Alcohol		1.00
3.	Stearic Acid (triple pressed)		1.50
4.	Cholesterol (USP)		0.60
5.	"Oat-Pro®"		2.00
6.	Lanolin (cosmetic anhydrous)		2.00
7.	Mineral Oil (65/75 saybolt)		12.00
8.	PVP K-30		5.00
9.	Triethanolamine		0.40
10.	Methyl Paraben		0.10
11.	Propyl Paraben		0.05
12.	"Dowicil" 200		0.10
13.	Perfume		q.s.
14.	Deionized Water	q.s.	100.00

Weigh and add 1, 2, 3, 4, 6, 7, and 11 into a container and begin heating and stirring. At the same time weigh and add 14 into a container and begin heating and stirring. Add 8 and stir until the material is completely dispersed, and the resultant dispersion is clear. Add the "Oat-Pro®" and continue stirring until complete dispersion occurs and no lumps can be felt or seen. Add 9 and 10 and bring the temperature to 70-73 C. Add the mineral oil-containing blend to the "Oat-Pro®"–water containing blend, when both are at 70-73 C. When all the mineral oil-containing blend has been added, begin cooling the batch. Cool to 40-43 C and add the remaining ingredients. Continue cooling to 30-35 C and package.

Shampoo Conditioner

FORMULA NO. 1

1.	Deionized Water	q.s.	100.00
2.	"Miranol" C_2M-SF (conc.)		15.00
3.	"Sipon" LT-6		8.00
4.	Propylene Glycol		5.00
5.	"Polymer JR" 125		1.25
6.	"Carbopol"-941		0.75
7.	"Sequesterene"-NA 2		0.05
8.	Formaldehyde Solution		0.10
9.	Hydrochloric Acid		0.50
10.	"Oat-Pro®"		2.00
11.	Perfume		q.s.

Weigh and add 1 into a container and begin stirring (use of an agitator equipped with a stirrer capable of imparting relatively high shearing stress should be used). Add 5 and stir, while beginning to heat. Heat to 70-73 C. When 5 is thoroughly dispersed and hydrated (no lumps should be seen or felt), add 6. When complete dispersion and hydration of 6 has occurred, add the "Oat-Pro®" (10). Stir until the "Oat-Pro®" is completely dispersed, at which time add the remaining ingredients except 8 and 11. Cool the batch, while stirring, to 40-43 C and add 8 and 11. Continue cooling to 25-30 C and package.

No. 2

"Ammonyx®" MO	20.0
"Deriphat" 170-C	16.0
"Onamer®" M	6.7
Distilled Water	q.s. to 100

	No. 3	No. 4
"Crodafos" SG	4.5	4.5
"Carsonol" SLES-2	45.0	45.0
"Carsamide" CA	2.0	2.0
"Crotein" SPA	2.0	0
Perfume, Preservatives, and Color	q.s.	q.s.
Distilled Water	46.5	48.5

Dissolve "Crotein" in water and then blend all ingredients by warming and agitating until uniform and clear. Adjust pH to 6.5-7.5 using phosphoric acid.

No. 5

Triethanolamine Lauryl Sulfate	15.0
"Ucare" Polymer LR-400	1.0
Lauric Diethanolamide (high purity)	2.0
Deionized Water, Dye, Perfume, Preservative	q.s.

Dissolve the preservative in hot water approx. 60 C (140 F). Quickly add the "Ucare" Polymer LR-400, and stir until the resin is dissolved. Hydration of the "Ucare" Polymer LR-400 should be achieved in about 40 min. Add the triethanolamine lauryl sulfate (40% solution), and mix well until all gel particles are dissolved. Melt the amide, and combine with the above mixture. Continue stirring until the shampoo is homogeneous. Check the pH, and, if it is less than 7.2, adjust with triethanolamine. Finally, add the desired perfume and color.

Creme Rinse

FORMULA	No. 1	No. 2
"Triton" CG-400	7.5	—
"Triton" CG-500	—	3.75
Cetyl Alcohol	1.0	1.0
Water	91.5	95.25
Colorants, Perfumes	← As required →	

These rinses should be diluted for use at the rate of one tablespoonful per 8 oz of water.

These formulations are prepared as follows. To obtain a product of proper viscosity and satisfactory resistance to stratification, the temperature must not be less than 80 C (176 F) during the processing, and the surfactant must be fused and thoroughly mixed before charging.

1. Charge the water and heat to 80 C (176 F) in a vessel equipped with agitation. Agitate *throughout the entire operation* at a slow speed to prevent entrapment of air bubbles.
2. Add the powdered cetyl alcohol and agitate slowly until it is melted.
3. Fuse "Triton" CG-400 or "Triton" CG-500 at 55–70 C (130–160 F) in a separate vessel (or in a hot room) and mix thoroughly before weighing.
4. Charge the "Triton" CG- surfactant *hot* to the water and cetyl alcohol and continue slow agitation at 80 C (176 F) at least 30 min to provide an emulsion of maximum stability.
5. Shut off the heat and continue slow stirring until the mixture has cooled to 30 C (86 F). During the cooling period, add the coloring agents and perfumes solubilized with "Triton" X-100.

Notes:

1. Maintain slow mixing at least until the formulation has cooled below its melting point.
2. Potassium chloride (0.1–1%) is often used as a viscosity modifier in creme rinse formulations. Although the effect of this additive on these formulations has not been examined, it is expected to be compatible.

No. 3

"Natrosol" 250 HHR (3% sol'n)	93.0
"Onamer®" M	6.7
Distilled Water	q.s. to 100

Baby Shampoo

FORMULA NO. 1

Sodium Laureth-12 Sulfate (60%)	35.50
"Lexaine" C (30%) (cocamidopropyl betaine)	12.00
"Natrosol" 250 HHR	0.60
"Lexgard" M	0.20
"Lexgard" P	0.10
"Bronopol"	0.04
Water, Dye, Perfume, q.s. to	100.00

Disperse "Natrosol" in cold water. Heat to 50 C and stir until completely hydrated. Add the remaining components and stir until everything is dissolved. Adjust pH to 6.5-7.0 with sodium hydroxide or phosphoric acid.

No. 2

"Miranol" C2M	15.00
"Tween" 80	5.14
"Pationic" TEA-LMS	3.43
"Pationic" 138C	3.43
"Clindrol" 200-O	3.00
Deionized Water	70.00

Warm to 70 C and mix slowly.

Egg Shampoo

FORMULA NO. 1

"Maprofix®" SP	20.0
Powdered Egg	0.2
Color (Yellow)	q.s.

Opacifier ("Morton" E-295)	0.2
Water	59.2
Perfume	q.s.
Salt	q.s.

Add egg to 10 parts of water to disperse egg before addition to batch. Add opacifier to 10 parts of water to disperse before addition to batch. Adjust pH to 7.0. Add sufficient salt for 1500 cps—after all components added.

No. 2

"Lakeway" 301-10	40
1 : 1 Coconut Diethanolamide	4.0
Propylene Glycol	1.0
Sodium Chloride	4.0
Ethylene Glycol Monostearate	3.0
Spray Dried Whole Egg	2.0
Formalin	0.1
Phosphoric Acid	q.s. to pH 6.8-7.2
Water	Balance
Perfume	q.s.

Heat water and "Lakeway" 301-10 to 65–70 C adding amide slowly while mixing. Mix in propylene glycol and sodium chloride. Add e.g. monostearate slowly, mix thoroughly. Add egg powder. Adjust pH. Cool slowly to room temperature. Add perfume and package.

Olive Oil Vitamin Shampoo

A	"Neo-Fat" 265	13.96
	"Lantrol"	0.50
	Olive Oil	0.50
	"Cerasynt" 840	5.00
	Vitamin E	0.02

B	Triethanolamine	2.19
	Glycerin	5.00
	Potassium Hydroxide Pellets	4.21
	Deionized Water	67.72
	"Klucel" MF (3% aqueous sol'n)	0.9

Dissolve "Klucel" in water. Add the other ingredients in B to this solution. Heat to 80 C. Combine ingredients in A and heat to 80 C. Add A to B with stirring. Cool to secure heavy viscosity.

Herbal Shampoo

"Lexaine" IBC-70	20.00
Sodium Chloride	0.10
"Lexein"-X250	1.00
"Bronopol"	0.02
FD & C Yellow #5—½%	0.80
FD & C Blue #1—½%	0.08
FD & C Red #4—½%	0.04
Perfume 0-115 (UOP)	0.50
Water	77.38

Charge water into a suitable making tank equipped with an agitator and having provisions for heating and cooling. Heat to 70-90 C with moderate agitation. Increase agitator speed and gradually add "Lexaine" IBC-70. When this is all added and completely dissolved, cool with moderate agitation to 40-45 C; then add and disperse balance of ingredients in the order shown. Cool to 25 C and fill.

Oily Hair Shampoo

A	Sodium Lauryl Sulfate (30%)	27.00
	Methyl Paraben	0.15
	Deionized Water	q.s.
B	"Monamide" 150LW	5.00
	Peg 6000 Distearate	1.50
C	Perfume, Color	q.s.

Blend A while heating to 60 C. Add B to A while heating to 75 C. Mix for 15 min and cool to 40 C. Add C and cool to 30 C.

Pet Shampoo

FORMULA NO. 1

"Lakeway" 301-10	25
1 : 1 Coconut Diethanolamide	5
Lanolin	1.0
Hexachlorophene	0.5
Sodium Chloride	2.0
Propylene Glycol	2.0
Ethylene Glycol Monostearate	2.0
Phosphoric Acid	q.s. to pH 6.8–7.0
Water	Balance
Perfume and Dye	q.s.

Blend at 65–70 C in order listed, starting with water first. Cool slowly and package.

NO. 2

(Protein)

"Maprofix" SP	22.000
Methyl Paraben	0.200
Formaldehyde (37%)	0.100
"Lanasan" CL	0.100
Distilled Water	74.540
Citric Acid	0.046
Sodium Chloride	2.800
FD & C Blue No. 1 (1% sol'n)	0.037
D & C Green No. 1 (1% sol'n)	0.177

No. 3

(White Coat)

"Maprofix" SP	25.000
Formaldehyde (37%)	0.100
Distilled Water	71.184
Citric Acid	0.016
Sodium Chloride	2.333
FD & C Blue No. 1 (1% sol'n)	0.300
"Grocor" 5220	1.000
D & C Red No. 19 (1% sol'n)	0.067

No. 4

(Coat Brightener)

"Maprofix" SP	22.000
Methyl Paraben	0.180
Propyl Paraben	0.020
Formaldehyde (37%)	0.100
"Lanasan" CL	0.100
Distilled Water	74.934
Citric Acid	0.046
Sodium Chloride	2.533
FD & C Blue No. 1 (1% sol'n)	0.030
Optical Brightener	0.020
FD & C Yellow No. 5 (1% sol'n)	0.037

Perfume Stick

Sodium Stearate	9.0
"Pluronic" F	9.0
"Onyxol" 345	3.0
Glycerin	10.0
Perfume Oil	7.0
SDA-40 Alcohol	23.5
Propylene Glycol	32.5
"Perma Kleer" 100	1.0
"Onamer" M	3.3
Distilled Water	1.7

"Roll On Perfume"

	FORMULA NO. 1	NO. 2	NO. 3
"Pluronic" F-127 Polyol	21.0	28.5	26.0
Ethanol (95%)	33.0	28.5	33.0
Water	36.0	33.0	31.0
Perfume Oil	10.0	10.0	10.0

Dissolve "Pluronic" F-127 polyol in a mixture of cold water (5–10 C) and alcohol. Add perfume oil and mix gently. Product is a clear, viscous liquid suitable for application from a "roll on" bottle.

Cologne

FORMULA NO. 1

(Stick)

Sodium Stearate	6.5
Glycerin	2.5
Distilled Water	3.5
"Carbowax" 600	6.0
"Onamer" M	3.3
Perfume	3.0
SDA-40 Alcohol	75.2

NO. 2

Perfume	5.0
Distilled Water	15.0
"Onamer" M	3.3
"Brij" 58	10.0
"Super Amide" LM	5.0
Sodium Stearate	9.0
Propylene Glycol	52.7

No. 3

(Emollient)

A	SDA-40 Alcohol (95%)	70.5
	Deionized Water	19.5
B	Dipropylene Glycol	5.0
	"Lantrol" AWS	5.0
	Perfume	q.s.

Combine alcohol and water and stir. Combine the dipropylene glycol, "Lantrol" AWS and perfume and mix thoroughly. Add B to A with constant stirring.

No. 4

"Crodafos" N.3 (neutral)	1.50
"Polychol" 5	3.00
"Polawax" A.31	1.25
"Novol"	2.00
Ethanol SDA-40 (95%)	46.00
Perfume between (1.5–5%)	q.s.
Propellant 11	1.50
Propellant 114	3.50
Water	41.25

Sachet

FORMULA No. 1

A	Water	51.80
	"Lexgard" M	0.15
	"Lexgard" P	0.05
	"Pluronic" L-35	2.00
B	"Lexemul" 561	21.00
	"Lexate" TA	10.00
	Cetyl Alcohol	5.00
	Perfume	10.00

Charge the water and other ingredients of A into a suitable making tank and heat to 70-75 C with stirring until Lexgards are dissolved. Weigh and melt the ingredients of B in a separate container; mix until homogeneous and bring to 70-75 C. Add B gradually to A with vigorous mixing and cool with stirring to 40-45 C after addition is complete. Add and thoroughly incorporate perfume. Package. Product sets to a firm heavy-bodied cream after 24 h at room temperature.

No. 2

"Tetronic" 1504	2
Propylene Glycol	6
Lanolin	2
Ceteareth-5	5
Glyceryl Monostearate	10
Perfume	4
Water	61
Preservative	q.s.

Add water, "Tetronic" polyol, propylene glycol, glyceryl monostearate, Ceteareth-5 and lanolin to mixing vessel. Warm gently to 60 C and mix in perfume and preservative. Transfer to containers. Product sets up into a viscous cream at room temperature.

No. 3

"Pluronic" 17R8	16
Propylene Glycol	6
Glyceryl Monostearate	10
Lanolin	2
Stearic Acid	3
Perfume Oil	4
Water	59
Preservative	q.s.

Add water, "Pluronic" polyol, propylene glycol, glyceryl stearate, stearic acid and lanolin to container. Warm gently to 60 C and mix well with a minimum of aeration, until mixture is homogeneous. Allow to cool to 45 C, add perfume oil and preservative, and mix well. Transfer to containers and allow to cool to room temperature at which it sets up into a stiff gelatinous cream.

Nail Conditioner

FORMULA NO. 1

"Syncrowax" ERL-C	3.00
"Syncrowax" HGL-C	10.00
"Fluilan"	6.00
Castor Oil	38.49
"Hartowax"	3.00
"Novol"	38.50
"Crotein" HKP	0.50
"Crotein" CAA	0.50
B.H.A.	0.01
Fragrance	q.s.

Melt all components together. Cool to just above setting point (45 C) and add the "Crotein" CAA and HKP with agitation and fill.

NO. 2

"Lanexol" AWS	1.5
"Crotein" SPO	0.5
Alcohol SDA-40 (95%)	53.0
Water	45.0

Dissolve the "Crotein" SPO in the water, add alcohol, "Lanexol" AWS, stir until uniform. Heat is not required. It is recommended that a lemon perfume be used in this system.

NO. 3

A	"Lexamine" S-13	2.00
	"Lexol" PG 8-10	2.00
	PEG 400 Distearate	1.50
	"Lexemul" 515	6.00
	Cetyl Alcohol N.F.	6.00

B "Lexein" A220	50.00
"Lexgard" M	0.20
"Bronopol"	0.01
Citric Acid Monohydrate	0.40
Water	31.89
Perfume	q.s.

Combine A in a vessel and heat to 65–70 C until dissolution. In a separate vessel, combine B and heat to 65–70 C with dissolution. Add A to B at 65–70 C with agitation. Cool batch, add perfume at 40 C and fill at 30 C.

Nail Hardener

Nitrocellulose (RS ¼ s)	12.5
"Cellosolve" Stearate	6.0
Tricresyl Phosphate	5.0
"Cellosolve"	40.0
Ethyl Acetate	17.5
Butyl Acetate	17.5
"Crotein" AD Anhydrous	2.4
Formalin	0.1

Dissolve nitrocellulose, tricresyl phosphate, cellosolve stearate in the cellosolve. Add "Crotein" to ethyl and butyl acetates and add to rest. When all dissolved, add formalin and bottle.

Nail Polish

Nitrocellulose CA4 E 32	14.2
"Santolite" MS 80	7.4
Camphor	1.9
Dibutyl Phthalate	4.8
Isopropanol	10.0
Butyl Acetate	25.4
Ethyl Acetate	6.5
Toluene	29.8

Nail Polish Remover

"Crotein" IP	0.1
"Crodamol" DA	2.5
"Lanexol" AWS	2.5
Methyl "Cellosolve"	94.9
Perfume	q.s.

Blend all ingredients, filter and fill.

Cuticle Remover

FORMULA NO. 1

"Pluronic" F-127 Polyol	22
Lactic Acid	5
Water	73
"Dowicil" 200	q.s.

Place water in container, cool to 5-10 C, add "Pluronic" F-127 polyol slowly while mixing. Maintain solution temperature at or below 10 C. When solution is complete, add lactic acid and preservative. Mix gently to insure homogeneity and transfer to containers. Product will set up into a crystal clear gel when it warms to room temperature. The viscosity of the final product may be altered by varying the amount of "Pluronic" F-127 polyol used.

NO. 2

A	"Veegum"	1.0
	Water	75.0
B	Triethanolamine	11.0
	Glycerin	10.0
C	"Brij" 30	1.2
	"Brij" 35	1.8
	Preservative	q.s.

Add the "Veegum" to the water slowly, agitating continually until smooth. Add B to A and heat to 70 C. Heat C to 75 C. Add to A and B and mix to 40 C.

NO. 3

"Poloxamer" 407	22
Lactic Acid	5
Water	73
Perfume, Preservative	q.s.

Cuticle Cream

1.	"Polawax"		7.50
2.	Stearyl Alcohol		6.00
3.	"Lantrol"		2.00
4.	Mineral Oil (65/75 saybolt)		15.00
5.	"Cetiol" V		4.00
6.	Isopropyl Palmitate		2.00
7.	Propylparaben		0.10
8.	Deionized Water	q.s.	100.00
9.	Methylparaben		0.10
10.	Glycerin		5.00
11.	"Oat-Pro®"		2.00
12.	Potassium Sorbate		0.10

Weigh the "oil" phase 1-7; begin heating and stirring. Heat to 70-73 C. Weigh the "water" phase 8-11 and begin heating and stirring. Heat to 70-73 C and add the water phase to the oil phase. Both should be 70-73 C. Cool to 40 C and add 12. Fill 25-30 C.

Rouge

FORMULA NO. 1

(Liquid)

White Mineral Oil	39.4
Oleic Acid	7.3

Dry Ingredients:

Titanium Dioxide	0.6
Zinc Stearate	0.42
Pigment	0.42
Aluminum Hydroxide	0.36

Water	47.7
Triethanolamine	3.7
Methyl p-Hydroxybenzoate	0.1
Perfume	q.s.

No. 2

A	Pigment grind	
	Color	1.25
	Titanium Dioxide	1.25
	Mineral Oil	2.50
	Preservative	As needed
B	Oleic Acid	6.00
	Propylene Glycol Monostearate	1.70
	Mineral Oil	25.00
C	Cellulose Gum, CMC-7LF	0.10
	Distilled Water	9.90
D	"Veegum"	0.75
	Distilled Water	14.25
E	Propylene Glycol	5.00
	Triethanolamine	3.00
	"Darvan" No. 1	0.25
	Distilled Water	29.05

1. Grind A in a roller mill until smooth.
2. Melt B and add Step 1. Heat to 75 C.
3. With C disperse cellulose gum in water.
4. With D disperse "Veegum" in water.
5. Add mixed ingredients from E to Step 3 mixture. Add Step 4 mixture and heat to 70 C.
6. Add Step 2 mixture to Step 5 mixture. Mix with high-speed agi-

tation. Cool with stirring in a cold-water bath.

Cream Lip Rouge

A	"Veegum"	0.5
	Sodium Carboxymethylcellulose (low visc.)	0.2
	Water	29.7
B	"Darvan" No. 1	0.3
	Propylene Glycol	5.0
	Water	30.6
C	"Nytal" 400	7.5
	Titanium Dioxide	1.0
	Color	q.s.
	Lanolin	6.0
	Petrolatum	6.0
	Isopropyl Myristate	3.0
	Beeswax	4.0
D	Carnauba Wax	2.0
	Oleic Acid	3.0
E	Morpholine	1.2
	Perfume and Preservative	q.s.

Dry blend the "Veegum" and CMC and add to the water slowly, agitating continually until smooth. Mix A and B. Heat to 65-70 C. Mill C, add to D and heat to 70 C. Add E to water phase, then immediately emulsify by adding oil phase with constant mixing until cool.

Skin Tint

FORMULA NO. 1

1.	Lantrol	0.3
2.	Stearic Acid (triple pressed)	3.25
3.	Polawax	1.5
4.	Mineral Oil (65/75 saybolt)	6.0
5.	Propylparaben	0.10

6.	Methylparaben		0.10
7.	"Oat-Pro®"		2.0
8.	Deionized Water	q.s.	100.00
9.	Propylene Glycol		4.0
10.	Titanium Dioxide		2.0
11.	Lo Micron Pink #25-11		0.7
12.	Yellow #2576		0.3
13.	Red #2513		0.1
14.	"Dowicil" 200		0.10
15.	Perfume		q.s.
16.	Triethanolamine		1.0

Weigh the oil phase 1-5; commence stirring and heating. Heat to 70-73 C. Weigh 6, 7, 8, and 9 and heat while stirring to 70-73 C. Add the "aqueous" phase to the oil phase. Continue stirring and add 10, 11, 12, and 13. Stir until a uniform dispersion results, add 16; cool to 35-40 C. Add 14 and 15. Fill at 25-30 C.

No. 2

1.	"Myvacet" Type 9-40		0.25
2.	Stearic Acid (triple pressed)		3.5
3.	Glyceryl Monostearate (non self-emulsifiable)		1.7
4.	Lanolin		2.0
5.	Mineral Oil (65/75 saybolt)		8.0
6.	Propylparaben		0.1
7.	Methylparaben		0.1
8.	"Oat-Pro®"		1.5
9.	Deionized Water	q.s.	100.00
10.	Propylene Glycol		3.0
11.	Titanium Dioxide		2.0
12.	Red #2513		0.9
13.	Ultra Blue 3585		0.2
14.	Triethanolamine		1.0
15.	Sorbic Acid		0.1
16.	Perfume		q.s.

Weigh 1-6 and heat, 70-73 C, while stirring continuously. Weigh 7-10 and heat to 70-73 C. Add this to the heated oil phase components (both phases should be at 70-73 C). Add 11, 12, and 13 and mix until uniformly dispersed. Add 14, cool to 35-40 C and add 15 and 16. Fill at 25-30 C.

Lip Gloss

FORMULA NO. 1

1.	Ozokerite Wax		4.00
2.	Lanolin (anhydrous)		3.50
3.	Ouricury Wax		5.50
4.	Wool Wax Alcohol		5.00
5.	"Ritaderm"		10.00
6.	Propylparaben		0.10
7.	BHA		0.05
8.	Bentone Gel IPM		5.00
9.	"Ritalin C"	q.s.	100.00
10.	Flavor		q.s.

Add 9 into a container and heat to 70-73 C, while stirring by means of an agitator equipped with a stirrer capable of imparting relatively high shearing stress (a propellor type is recommended). Add 8, and stir until the dispersion is homogeneous. Add the remaining ingredients, with the exception of 10, while maintaining the temperature at 70-73 C. When all the ingredients have completely melted, and the resultant dispersion is uniform, cool to about 60 C, and add 10. Package in suitable containers.

NO. 2

Carnauba Wax	3.6
"Lantrol"	8.0
Propyl Paraben	0.1
"Acetol"	3.0
Ozokerite (170 F)	7.0
Candelilla Wax	6.0
Camphor	0.1
Castor Oil	72.2

Heat to 70-80 C. Mix slowly and pour into molds.

No. 3

"Syncrowax" HR-C	5.0
"Syncrowax" HGL-C	5.0
"Timica" Silk White	8.0
"Fluilan"	8.0
Color	1.0
"Kaydol"	73.0

Melt waxes. Add "Fluilan" and rest of "Kaydol." Disperse color blend and "Timica" into this mixture under low shear. Deaerate at 65 C in water bath. Fill off at 50 C.

Lip Balm Sunscreen

"Amerscreen" P	1.0
"Acetulan"	4.0
"Amerlate" P	10.0
Isopropyl Palmitate	11.0
Beeswax (USP)	9.0
Ozokerite	5.0
Candelilla Wax	7.0
Carnauba Wax	4.0
Castor Oil	49.0
Perfume and Preservative	q.s.

Heat all ingredients to 90 C until uniform. Mix while cooling to 75 C. Mold.

Lipstick

FORMULA No. 1

"Lanocerin"	8.5
Lanolin	12.5
Castor Oil	9.0
Cetyl Stearyl Alcohol	5.0
Candelilla Wax	13.5
Myristyl Myristate	5.0

"Isopropylan" 50	15.0
Color Pigments (35% ground in "Isopropylan" 50)	18.0
Synthetic Pearl (70% castor oil)	13.5

Pregrind colors in "Isopropylan" 50 plus some castor oil. Add synthetic pearl next, then add this color/pearl mixture to the balance of the formula. Pour at 70 C.

No. 2

Lanolin	12.5
Castor Oil	15.0
Beeswax (USP)	5.0
"Lanocerin"	8.5
"Isopropylan" 33	7.5
"Isopropylan" 50	7.5
Myristyl Myristate	5.0
Candelilla Wax	13.5
"Timica" Pearlwhite	13.5
Pigment (35% castor oil)	12.0

No. 3

Ozokerite	3.50
Carnauba Wax	4.00
Candelilla Wax	5.00
Spermwax	7.00
"Ritalan"	39.66
"Ritaderm"	15.00
Propyl Paraben	0.10
BHA	0.10
D & C Red No. 7	0.30
D & C Red No. 9	0.55
Titanium Dioxide	0.50
Castor Oil	q.s.

Weigh and add the ozokerite, carnauba wax, candelilla wax, spermwax, "Ritalan," "Ritaderm," propyl paraben and BHA into a container, stirring continuously while heating to 75-80 C. (An agitator equipped with a stirrer capable of imparting relatively high shearing stress is recommended.) Weigh and add the D & C Red No. 7, D & C Red No. 9, titanium dioxide and castor oil into another container and stir until completely dispersed. Then add the second blend to the first and cool, while stirring, to 68–70 C. Pour into suitable molds.

No. 4

A	1.	Candelilla Wax		5.00
	2.	Carnauba Wax		2.00
	3.	Ozocerite		1.50
	4.	"Emerwax" 4266-0		1.50
	5.	"Cetiol" V		4.00
	6.	Mineral Oil (65/75 saybolt)		5.00
	7.	Beeswax		8.00
	8.	Lanolin (cosmetic grade)		10.00
	9.	Castor Oil	q.s.	100.00
	10.	"Oat-Pro®"		2.00
	11.	Color Pigments		47.00
B	1.	Titanium Dioxide		8.00
	2.	D & C Red #19		2.25
	3.	D & C Red #21		1.50
	4.	Castor Oil		35.25

Weigh A (1–9) and begin heating and stirring. Heat to approximately 80 C and mix until all waxes have melted. Prepare B several hours in advance by adding 1–3 to the castor oil and processing with a roller mill. Cool A to 70–73 C and add 10 and B (11). Mix thoroughly and pour into molds.

No. 5

"Lexemul" 515	42.00
Refined Castor Oil	36.80
Mineral Oil	8.00
Bromo Acid	5.00
Petrolatum	4.00
Carnauba Wax	4.00
"Lexgard" P	0.20

Warm to 65 C and mix. Stir slowly while cooling.

	NO. 6	NO. 7
"Crodesta" F50	—	5.0
"Crodesta" S10	6.5	—
"Crodesta" A10	4.0	—
Carnauba Wax	2.0	3.0
Ceresine Wax	—	8.0
Candelilla Wax	12.0	—
Lanolin USP	10.5	5.0
Paraffin Wax 140	—	5.0
Microcrystalline Wax	—	5.0
Castor Oil	42.5	10.0
"Crodacol" C	—	5.0
"Novol"	—	34.0
"Crodamol" IPM	2.0	—
2 Ethylhexyl Palmitate	—	20.0
"Robane"	6.5	—
Pigments Ground in Castor Oil (35% colors)	14.0	q.s.
Perfume, Preservative, Antioxidant	q.s.	q.s.

Melt all the ingredients together. Add the pigment paste with gentle stirring, allow to deaerate, then pour into molds and cool.

Mud Pack

1.	Sorbitol		5.00
2.	Deionized Water	q.s.	100.00
3.	Methylparaben		0.10
4.	Fullers Earth		30.00
5.	"Oat-Pro®"		5.00
6.	Propyl Paraben		0.075

Weigh 1–3 and 6 into a container and commence stirring. Add 4 and 5 and stir until a smooth homogeneous paste results.

Face Mask

FORMULA NO. 1

1.	Deionized Water	q.s.	100.00
2.	"Veegum"		10.00
3.	"Oat-Pro®"		4.00
4.	Methylparaben		0.10
5.	Propyl Paraben		0.10
6.	Ethyl Alcohol (SDA-40)		18.00
7.	Color		q.s.
8.	Perfume		q.s.

Weigh 1 and begin stirring. Add 2 and mix until the resultant dispersion is smooth and lump free. Add 3-8. Continue mixing until the dispersion is smooth.

NO. 2

(Peelable)

1.	"Oat-Pro®"		5.00
2.	"Gelvatol" 3/90		10.00
3.	Deionized Water	q.s.	100.00
4.	Glycerin		1.00
5.	"Resyn" 2260		5.00
6.	Ethyl Alchohol SDA-40		16.00
7.	2-Amino-2-Methyl–1-3 Propanediol		0.02
8.	Color		q.s.
9.	Perfume		q.s.

Weigh 3 and add while stirring continuously, 1 and 2. Heat to 70-73 C, begin cooling after mixing for 15 min at this temperature. Continue stirring. Add 4 and 5 at 50-54 C. Add 6. Continue stirring and cooling and add 7-9 at 25-30 C.

NO. 3

"Pellulan" (molecular weight 150,000)	20.0
Cellulose Gum	5.0
Glycerin	2.0

Ethyl Alcohol	5.0
Distilled Water	67.0
Preservative	As needed
Perfume	As needed

No. 4

Bentonite	12.0
TiO_2	6.0
Cellosize (low visc.)	0.5-1.0
Ethanol SDA-40	10.0
Water	q.s.
Certified Colors	q.s.
Methyl Paraben (USP)	0.2
Perfume Solubilized with "Volpo" 10	q.s.

No. 5

"Albagel" (natural)	17.0
Ethanol SDA-40	10.0
TiO_2	5.0
"Methocel" 60 HG	0.5-1.0
Water	q.s.
Perfume Solubilized with "Volpo" 10	q.s.
Methyl Paraben (USP)	0.2

Micropulverize the TiO_2 and mineral gum. Dissolve perfume in "Volpo" 10 and alcohol. Disperse the organic gum in the water and sprinkle in the "Albagel"/TiO_2 blend under high speed kitchen-aid type mixing. When smooth, add alcohol phase and continue mixing until alcohol phase is well blended.

No. 6

(Acid)

| A | "Veegum" K | 6 |
| | Water | 83 |

B	Ethanol		4
C	Glycerin		4
	Sulfated Castor Oil		3
	Color and Preservative		q.s.
	Buffered to pH 5.5		

Buffer solution: 60 parts—1 M citric acid,
40 parts—saturated sodium citrate solution.

Talcum, Body

FORMULA NO. 1

1.	Talc	q.s.	100.0
2.	Magnesium Stearate		2.5
3.	Zinc Oxide		2.0
4.	"Oat-Pro®"		3.0
5.	Perfume		q.s.

Add 2-5 to 1 and blend until completely uniform.

NO. 2

(Quick-Breaking Foam)

Talc	25.000
"Emcol" E-607S	1.200
"Emcol" E-607L	0.200
"Witconol" F26-46	0.250
Water	31.675
Ethyl Alcohol	31.675
Propellants 12/114 (20 : 80)	10.000

Mix water and ethyl alcohol, then stir in surfactants. Add talc and stir until a uniform dispersion is effected. Transfer to aerosol containers, crimp and pressurize.

The above formulation can be pigmented for makeup applications. The quick rub-in of the talc with uniform spreading yields a lubricious, nondusting film which makes this formulation suitable for pre-electric shave, surgical glove lubricant and nondusting body or baby talc.

Baby Powder

1.	Talc	90.15
2.	Zinc Stearate	2.50
3.	Syloid #72	2.00
4.	"Oat-Pro®"	3.00
5.	Zinc Oxide	3.00
6.	Dioxin	2.00
7.	Perfume	q.s.

Add ingredients 2–7 to the talc; blend until completely uniform.

Toilet Soaps

FORMULA NO. 1

1.	Olive Oil (refined)	200.00
2.	Stearic Acid (triple pressed)	100.00
3.	Corn Oil	100.00
4.	Oleic Acid	100.00
5.	Butylated Hydroxyanisole	0.15
6.	Sodium Hydroxide (USP)	100.00
7.	Deionized Water	100.00
8.	Deionized Water	650.00
9.	"Oat-Pro®"	150.00
10.	Perfume	q.s.
11.	Color	q.s.

Heat 8 to 75 ± 3 C and sprinkle in 9 while mixing with a high shear propeller type stirrer. Heat 1–5 to 92 ± 2 C while stirring slowly to prevent air entrapment. Dissolve 6 and 7. While maintaining the "oil" mixture at 92 ± 2 C slowly add aliquots of the "Oat-Pro®" dispersion and sodium hydroxide. Maintain the temperature for 1 h after completing the addition of "Oat-Pro®" and sodium hydroxide while continuing to slowly stir the batch. Cool to 68 ± 2 C and add 10 and 11. Continue cooling to about 25 C. This product can be processed, to remove excess moisture, by applying a vacuum during the compounding operation. The product can also be warmed to about 45 C, for 24 h prior to pressing to facilitate moisture removal.

No. 2

"Tetronic" 1508 Polyol	18.00
"Pluracol" E-4000 Polyol	10.00
Soap (20 coco, 80 tallow)	40.00
Hydrogenated Peanut Oil	10.00
Super-Pro 5A	5.00
Sodium Lauroylisethionate	5.40
Stearic Acid (triple pressed)	5.00
Hydrogenated Cottonseed Oil	5.00
Titanium Dioxide	1.00
Perfume	0.60

Mix all ingredients in a suitable container. Pass over a nine roll mill and convert into ribbons. Feed ribbons to a double vacuum plodder and extrude at 60–65 C. Cut and press in suitable equipment. Mill and plodder may require more power than similar equipment used for soap.

No. 3

1.	Bioterge AS 90		30.00 .
2.	"Carbowax" 6000		20.00
3.	"Carbowax" 400	q.s.	100.00
4.	"Oat-Pro®"		25.00
5.	"Ritalan"		1.00
6.	Citric Acid		0.40

Add the "Carbowax" 6000 into a container and begin heating and stirring. Heat to 85–90 C (until the "Carbowax" 6000 melts). Add the "Carbowax" 400 and continue stirring. Sprinkle in 4 and continue mixing until a uniform slurry results. Add 5, maintain the temperature at 80–85 C and continue mixing. Add 1 and 6 and begin cooling. Cool the batch to 45 C and micropulverize the blend. Press into cakes of desired size and shape.

No. 4

1.	Igepon AC 78	30.50
2.	Stearic Acid (triple pressed)	26.00
3.	Glycerin (USP)	17.00
4.	Deionized Water	10.00
5.	"Oat-Pro®"	15.00
6.	Citric Acid	0.50
7.	Methyl Paraben	0.05
8.	Propyl Paraben	0.05
9.	"Irgasan" DP300	0.10

Weigh and add 1, 3, and 4 into a container and begin stirring and heating. Heat the blend to 85-90 C. Add 2, 5, 6, 7, and 8 and continue stirring. Cool the batch to 40-45 C. Add 9 and micropulverize. Press into cakes of desired size and shape.

Glycerin Gel

"Poloxamer" 407	17
Glycerin (95%)	35
Water	48
Preservative	q.s.

Hydrogen Peroxide Gel

"Pluronic" F-127 Polyol	25
Hydrogen Peroxide (30% sol'n)	10
Water (deionized or distilled)	65

Cool the deionized water to 5-10 C and place it in a mixing container. Add the "Pluronic" F-127 polyol slowly with good agitation and continue mixing until solution is complete, maintaining the temperature below 10 C. Add the cool hydrogen peroxide solution (5-10 C) slowly with gentle mixing. Immediately transfer to containers and allow to slowly warm to room temperature whereupon the liquid becomes a crystal clear ringing gel. An alternate procedure is to combine the cool hydrogen peroxide and the cool water followed by the addition of the F-127 to the aqueous system with mixing. Maintain a 5-10 C temperature until all the "Pluronic" polyol has gone into solution. Bottle immediately.

Phytocosmetic Emulsion

Stearic Acid	3.00
Glycerol Monostearate	3.00
Vegetable Oil	10.00
Squalane	10.00
Soya Sterol	3.00
(Base)	q.s.
Water	to 100

Mascara

FORMULA NO. 1

(Waterproof)

A	Carnauba Wax	1.1
	"Emerwax" 1253	8.7
	Ozokerite (170 F)	6.5
	"Emersol" 132	2.2
	Propyl Paraben	0.2
	"Trisolan" LP	2.2
B	Deionized Water	55.0
	"Veegum" HV	0.5
	Cellulose Gum	0.75
	Methyl Paraben	0.25
	Cosmetic Black	1.1
	Channel Black	1.1
	"Carboset" 514	13.2
C	Morpholine	1.7
	Deionized Water	5.5

1. Mix ingredients in A and heat to 85 C.
2. Mix water, "Veegum," cellulose gum, and methyl paraben, and heat with rapid agitation to 85 C.
3. Slowly add pigments from B to Step 2. Recirculate through a colloid mill or homogenizer-tight gap and retare for water loss.
4. Mix Step 1 into Step 3.

5. Follow immediately with C mixture.
6. Cool to 35 C. Fill into mascara cases or bottles.

No. 2

(Cream)

A	"Veegum"	2.0
	Sodium Carboxymethycellulose (low visc.)	0.1
	Water	47.9
B	"Darvan" No. 1	0.2
	Propylene Glycol	1.5
	Water	30.9
C	Beeswax	6.5
	Light Mineral Oil	3.5
	Channel Black	1.0
D	Stearic Acid	1.0
	Carnauba Wax	5.0
E	Morpholine	0.4
	Preservative	q.s.

Dry blend the "Veegum" and CMC and add to the water slowly, agitating continually until smooth. Mix A and B and heat to 65-70 C. Mill C, add to D and heat to 70 C. Add E to the A and B mixture and then immediately emulsify by adding the C and D mixture, constantly mixing until cool.

No. 3

1.	Beeswax (USP white)		30.00
2.	Mineral Spirits		52.00
3.	Propyl Paraben		0.10
4.	Pure Oxy Black 3068		5.00
5.	Deionized Water	q.s.	100.00
6.	Borax		0.60
7.	"Oat-Pro®"		2.00
8.	Bronopol		0.04

Weigh and add 1 and 3 into a container and begin heating and stirring. Heat the blend to 75–80 C and maintain at this temperature. Weigh and add 5 into another container and begin stirring and heating. Use of a relatively high shear imparting stirrer is recommended. (A variable speed agitator equipped with a propeller type stirrer is suggested). Add the "Oat-Pro®" while continuing to heat the batch. When the "Oat-Pro®" is completely dispersed, add 6 and bring the temperature to 75–80 C. Add the "Oat-Pro®"-containing blend, which is at 75–80 C to the beeswax-containing blend, which is at the same temperature. When all the "Oat-Pro®" blend has been added, add the remaining ingredients, *with the exception of* 8. Begin cooling; cool to 50 C. Continue stirring, while adding 8 at 50 C. Then package.

No. 4

(Liquid)

A	"Emerwax" 1253	10.00
	Ozokerite 170-D	10.00
	Cetyl Alcohol, N.F.	8.00
	Carnauba Wax	2.00
	"Lanfrax"	0.50
B	Mineral Spirits	59.50
C	Pigments	10.00

Dissolve all the waxes in the mineral spirits at 90 C. Employ a reflux column. Allow the mixture to cool to 50 C. Add pigments under high speed agitation.

Kohl Stick

Vegetable Oil	48.0
Carbon Black	13.0
Anatase (TiO_2)	6.0
Carnauba Wax	4.0
Ozokerite 170	6.0
Beeswax	13.0

Isopropyl Myristate	9.8
Propyl Paraben	0.1
BHA	0.1

Eye Shadow

FORMULA NO. 1

(Pearlescent)

A	"Veegum" F	5
	"Mearlin" AC	35
	"Nytal" 400	29
	Zinc Stearate	8
	Magnesium Carbonate	1
B	Acetol	3
	Polysorbate 20	9
	Water	10
	Preservative	q.s.

Micropulverize A. Mix B and add to A. Continue to blend in mixer. Screen through a No. 16 sieve. Compress.

NO. 2

(Frosted)

French Talc	18.4
Italian Talc	18.4
Zinc Stearate	5.0
Kaolin	5.0
Ultramarine Blue	4.0
Hyd. Chrome Green Oxide	4.0
"Preservatol"	0.2
Bismuth Oxychloride	40.0
"Ceraphyl" 375	5.0

Premix dry ingredients in a ribbon-blender. Then add "Ceraphyl" 375 by spraying while powder is blending. Pass through micropulverizer with 0.027 in. herringbone screen. Mixture is now ready for pressing.

No. 3

Italian Talc	33.00–37.00
Zinc Stearate (USP)	2.00
Colloidal Kaolin	5.00
Pigments	q.s.
Bismuth Oxychloride	30.00
Titanium Dioxide and Mica	20.00
"Emerest" 2310	6.00–10.00

Pulverize the first four ingredients until uniform. Mix in the "Emerest" 2310 and gently blend in the bismuth oxychloride and mica. Press into pans.

	No. 4	No. 5	No. 6
Mineral Oil	41.00	9.50	26.00
Ozokerite 170	23.80	36.00	22.00
Beeswax	15.90	18.00	13.00
Cetyl Alcohol	9.20	.15	–
Candelilla Wax	7.20	–	4.00
Butyl Stearate	2.90	–	–
Cocoa Butter	–	19.00	–
Petrolatum	–	5.20	18.00
Oleyl Alcohol	–	3.00	3.00
Cholesterol	–	.15	–
Pigment/Pearl	q.s.	9.00	14.00

No. 7

A	Ultramarine Blue #3516	5.55
	Zinc Stearate (USP)	3.70
	Kaolin #347	11.00
	Mica #280	40.00
	"Avicel" PH-105 MCC	32.15
B	"Schercomid" AME-70	7.60

A–Premix all powders in PK blender for 10 min.

B—Add AME-70 through spinbar and mix for 5 min. Load mix to compression die and press at 2800 psig.

A alone does not press at all. At 75% of the above AME-70 level of B, the product shows "soft press" features—soft surface and slight cracking when dropped 3 ft onto a concrete surface. When the AME-70 content reaches 7.5–8% level in the mix, a good hard surface results. No cracking occurs when dropped 3 ft onto a concrete floor when this formulation is pressed at 2800 psig with 10–15 s dwell time.

NO. 8

(Powder)

"Veegum" F	7
"Nytal" 400	50
Zinc Oxide	4
Zinc Stearate	11
Kaolin	10
Pigments	18

NO. 9

(Eye "Cover")

"Croda" Liquid Base	30.0
"Super Corona" Lanolin (USP)	3.0
Microcrystalline Wax	1.0
Carnauba Wax	4.5
"Novol"	10.0
"Provol" 10	10.0
Propyl Paraben USP	0.2
Magnesium Carbonate (light USP)	3.5
TiO_2 (328 grade)	32.5
Talc	4.0
Red Iron Oxide	0.12
Yellow Ochre	0.88
Perfume	0.3

Homomixer or similar required. Melt the oil phase to 75 C and begin homogenizing. Dust in well blended pigment phase, holding temperature at 75 C for 20-30 min. Add perfume and stir briefly. Transfer to "stay warm" filler and fill at 65-75 C.

NO. 10

A	Methyl Paraben	0.15
	Glycerol Monooleate	2.00
	"Neo-Fat"	3.00
	Ethylene Glycol Monostearate	4.00
	Isopropyl Myristate	9.00
B	"PVP" K-30	1.00
	Triethanolamine	1.00
C	Titanium Diomide	5.50
	Talc	2.50
	Bismuth Pearl	2.00
	Sodium Lauryl Ether Sulfate	1.00
	Pigments	q.s.
	Deionized Water	q.s. to 100

NO. 11

1.	Mineral Oil (65/75 saybolt)		8.0
2.	"Amerlate" P		2.0
3.	Stearic Acid (triple pressed)		4.0
4.	Glyceryl Monostearate (non self emulsifying)		5.0
5.	Cetyl Alcohol		2.0
6.	"Cetiol" V		5.0
7.	Propylparaben		0.10
8.	Methylparaben		0.10
9.	"Oat-Pro®"		1.0
10.	Triethanolamine		1.0
11.	Propylene Glycol		5.0
12.	Sodium Benzoate		0.10
13.	Deionized Water	q.s.	100.00

Pigments:

1.	Chromalite Dark Blue	3.5
2.	Chromalite Magenta	2.0
3.	Pearl Glow	5.0

Weigh the "oil" phase ingredients (1-7), begin heating and stirring. Heat to 70-73 C. Weigh 13, 11, 9, and 8 into another container and begin stirring and heating. Heat to 70-73 C. Add the aqueous phase components which are at 70-73 C to the oil phase components. Add the pigments and mix until the pigments are uniformly blended. Add 10 and cool to 35-40 C at which temperature add 12. Fill at 25-30 C.

Eye Cream

1.	Mineral Oil		6.00
2.	Lanolin AAA	q.s.	100.00
3.	Petrolatum (USP white)		42.90
4.	Propyl Paraben		0.10
5.	BHA		0.10
6.	"Oat-Pro®"		2.00

Weigh and add all the ingredients, with the exception of the "Oat-Pro®," into a container. Begin heating and stirring. Heat to 70-73 C and stir by means of an agitator equipped with a propeller type stirrer. Slowly add the "Oat-Pro®" and stir until complete dispersion has occurred. Begin cooling the batch; cool the batch to 48-51 C and package at this temperature, while continuing to stir.

Eye Highlighter

1.	"Ritachol"		3.0
2.	Isopropyl Myristate		1.0
3.	Stearic Acid (triple pressed)		2.0
4.	Glycerol Monostearate (self-emulsifying)		1.0
5.	Propyl Paraben		0.10
6.	Methyl Paraben		0.10
7.	Ultra Blue #3516		0.30
8.	Deionized Water	q.s.	100.00
9.	"Carbopol" 940		0.35
10.	Propylene Glycol		3.0
11.	"Carbowax" 400		3.0
12.	Triethanolamine		1.00

13.	"Bilite" Ultrawhite	15.00
14.	"Oat-Pro®"	5.00
15.	"Dowicil" 200	0.10

Weigh and add 1-7 into a container. Begin heating and stirring. Heat the blend to 70-73 C. Weigh and add 8 into a container and begin stirring. Use a variable speed agitator capable of imparting high shear stress is recommended. A propeller type stirrer is suggested. Add 9 and stir until the material is completely dispersed and hydrated (no lumps of undissolved "Carbopol" should be seen or felt). Add 14 and stir until the "Oat-Pro®" is completely dispersed and hydrated. Begin heating and add the ingredients remaining with the exception of 12 and 15. Heat to 70-73 C. When both blends are at 70-73 C add the 1-7 blend to the "Carbopol" 940-containing dispersion. Add 12 and modify agitation as necessary due to the apparent viscosity of the formula. Cool to 43-45 C and add 15. Cool to 25-30 C and package.

Eyeliner

FORMULA NO. 1

"Pluronic" F-68	10-12
Talc	q.s.
Magnesium Stearate	3-10
Calcium Carbonate	8-10
Pigments	5-30

NO. 2

A	"Veegum"	2.5
	Water	75.5
B	Polyvinylpyrrolidone	2.0
	Water	10.0
C	Pigment	10.0
	Preservative	q.s.

Add the "Veegum" to the water slowly, agitating continually until smooth. Dissolve the polyvinylpyrrolidone in water, using a little heat.

Add B to A and then add C. Mix well.

Note: For a product with a soft, cream-like viscosity that can be applied with a brush, increase the "Veegum" content to 3.5%.

Eye Makeup Remover

FORMULA NO. 1

"Acetulan"	4.0
Mineral Oil (70% visc.)	62.0
Petrolatum (USP white)	12.0
Paraffin Wax (133 F m.p.)	4.0
Carnauba Wax	4.0
Ozokerite	10.0
Myristyl Lactate	4.0
Perfume and Preservative	q.s.

NO. 2

A	TEA Lauroyl Lactylate		15.0
	Sodium Isostearoyl-2-Lactylate		3.0
	Lauramide DEA		4.0
B	Methyl Paraben		0.2
	Deionized Water		77.8
	pH 7.6	Cloud Point < -2 C	

Eye "Lighter"

1.	Micro Crystalline Wax		3.00
2.	Ozokerite Wax		5.00
3.	Carnauba Wax		3.50
4.	"Cetiol" V		4.00
5.	Lanolin (anhydrous)		3.00
6.	Castor Oil	q.s.	100.00
7.	Titanium Dioxide		5.70
8.	Talc		4.10
9.	"Flamenco" Pearl SEC		10.00

10.	"Bi-Lite" Ultra Pearl	10.00
11.	BHA	0.10
12.	Propyl Paraben	0.10
13.	"Oat-Pro®"	2.00
14.	Silicone 344	5.00

Weigh and add 6, 7, and 8 into a container and begin stirring. Stir until the resultant dispersion is uniform and pass twice over a three roll mill. Weigh and add items 1-5 into another container and begin stirring and heating. Heat to 80–86 C while using an agitator equipped with a stirrer capable of imparting relatively high shearing stress. After all the ingredients which are solids are at room temperature have melted, begin cooling. Cool to 65-70 C and add the blend of 6-8. Continue stirring and cooling. Add the remaining ingredients at 53-55 C and continue to stir until the batch is homogeneous. Package at 48-52 C.

Pre-Electric Shave

FORMULA NO. 1

Alcohol SDA-40		75.00
Isostearyl Lactate		10.00
Cetyl Lactate		4.00
Deionized Water		11.00
Perfume		q.s.
Color		q.s.

Dissolve the isostearyl lactate and cetyl lactate in the alcohol. Add perfume. Slowly add water with agitation. Filter and color if desired.

NO. 2

A	"Veegum" T	1.8
	CMC 7M	0.1
	Water	38.0
B	Alcohol SDA-40	50.0
	Menthol	0.1
	Oleic Acid	1.0

Oleyl Alcohol	2.0
Propylene Glycol	1.0
"Laneto"-50	1.0
"Silicone Fluid" 200	1.0
"Tween" 80	2.0
Fumed Silica	2.0
Perfume	q.s.
Preservative	q.s.

Dry blend the "Veegum" T and CMC and add to the water slowly, agitating continually until smooth. Mix the ingredients in B until uniform. Add B to A slowly and mix until smooth. This formula is a soft cream and may be dispensed from a wide mouth jar or a plastic squeeze bottle.

No. 3

"Arlamol" E	10.0–15.0
Alcohol SDA-40	84.6–89.6
Menthol USP	0.2
Perfume	0.2

No. 4

"Luviskol" K30 Resin	1.0
Alcohol	80.0
Diisopropyl Adipate	5.0
Citric Acid	0.2
Water	13.8

Add alcohol to water, add citric acid, diisopropyl adipate and "Luviskol" resin in order, and mix gently until homogeneous. Add dye and perfume to suit. Product is a fluid liquid.

No. 5

"Pluronic" L-121 Polyol	2
Alcohol	60
Water	15
Isopropyl Myristate	19

"Laneth"-10 Acetate	4
Color	q.s.
Perfume	q.s.
Preservative	q.s.

Dissolve isopropyl myristate in alcohol. Add water, "Laneth"-10 acetate and "Pluronic" polyol and mix gently until homogeneous. Add color and perfume and transfer to suitable containers. Product is a clear, fluid liquid.

Pre-Shave Talc Stick

A	"Veegum"	1.9
	Water	q.s.
B	Zinc Stearate	4.7
	Light Magnesium Carbonate	1.9
	"Nytal" 300	91.5

Add the "Veegum" to about 30 parts of water slowly, agitating continually until smooth. Blend B. Add A to B and work into a smooth paste (add more water if necessary). Pack in stick mold and allow to dry.

After-Shave Applications

FORMULA NO. 1

Alcohol	13.5
Menthol	0.1
Zinc or Aluminum Phenolsulfonate	2.0
Perfume	0.5
"Tween" 20	1.9
Water	82.0

Dissolve the phenolsulfonate salt in water. Add "Tween." Dissolve the menthol in the alcohol. Mix the two solutions. Add the perfume slowly and with continuous stirring. Age and filter.

No. 2

Alcohol	40.0
Propylene Glycol	5.0
"Cremophor" RH-40 Dispersant	0.5
"Luviskol" K30 Resin	0.5
Water	54.0

Add alcohol to water, add propylene glycol, "Cremophor" dispersant and "Luviskol" resin in order and mix gently until homogeneous. Add dye and perfume to suit. Product is a fluid liquid.

No. 3

"Lanoquat" DES 25	0.40
Alcohol SDA-40 (anhydrous)	35.00
Deionized Water	64.40
Perfume	0.20

Separately add the perfume to the alcohol and the "Lanoquat" DES to the water. Add the alcohol mixture to the water mixture with mixing. Chill and filter.

No. 4

A	"Veegum"	1.00
	Water	82.80
B	Sodium Lauryl Sulfate	1.00
	Glycerin	2.00
	Allantoin	0.10
C	Polyethylene Glycol 400	4.00
	Isopropyl Myristate	1.00
	Acetulan	2.00
	Lecithin	1.00
D	Ethanol	5.00
	Menthol	0.10
	Preservative	q.s.

Add the "Veegum" to the water slowly, agitating continually until smooth. Add B to A and heat to 70 C. Heat C to 75 C. Add C to A and B with mixing until temperature drops to 40 C. Add D and continue mixing until cool.

No. 5

(Stick)

"Procetyl" AWS	5.0
Sodium Stearate	6.0
Silicone Fluid 470	2.0
Propylene Glycol (USP)	4.0
Ethanol SDA-40 (95%)	70.5
"Crotein" AD-X	1.0
Water	1.5
Perfume	10.0

Weigh all ingredients except perfume into a suitable jacketed mixer equipped with an efficient reflux condenser. Bring to about 75 C with agitation. When dissolved, begin cooling and add perfume about 55 C. Pour into molds at 50 C.

No. 6

(Pearly)

"Solulan" 98	3.00
"Mearlin" AC	0.10
"Natrosol" 250HR	0.10
"Carbopol" 941	0.15
Distilled Water	43.15
"Ceraphyl" 230	2.00
Alcohol SDA-40	50.00
Triethanolamine (10% sol'n)	1.50
Perfumes, Dyes, and Preservatives	q.s.

Blend "Solulan" 98 with "Mearlin" AC. Heat part of the water to 80 C and blend in "Natrosol." Disperse "Carbopol" in the remaining water and add "Natrosol" solution. Add "Ceraphyl" 230 dissolved in alcohol. Add triethanolamine solution, then the "Solulan" 98/"Meralin" AC blend.

No. 7

(Quick-Breaking Foam)

PEG-400 Monostearate	2.0
Capyrium Chloride	1.0
Cetyl Alcohol	6.0
Alcohol	54.6
Perfume	q.s.
Water	26.4
Propellant 12/114 (40/60)	10.0

Mix first five ingredients together until dissolved. Transfer to aerosol containers, add water, crimp valves, and pressurize.

Shave Cream

FORMULA No. 1

(Aerosol)

1.	Distilled Coconut Fatty Acid		1.320
2.	Low Titer White Oleic Acid		1.320
3.	Single Pressed Stearic Acid		1.730
4.	Eutectic Palmited Stearic Acid		3.080
5.	Stearic Acid (T.P.)		0.920
6.	Methyl Paraben		0.023
7.	Propyl Paraben		0.023
8.	"Tenox" B.H.A.		0.023
9.	Phosphoric Acid (85%)		0.591
10.	Sodium Hydroxide Sol'n. (75%)		6.000
11.	Deionized Water	q.s.	100.000
12.	Propylene Glycol		5.250
13.	"Oat-Pro®"		1.000
14.	"Dowicil" 200		0.023
15.	Perfume		q.s.

Weigh 11 into a container and begin stirring (a high shear imparting variable speed agitator is suggested). Add 13 ("Oat-Pro®") and stir until it is completely dispersed and hydrated (no lumps of "Oat-Pro®" should be evident). Add 12 to the "Oat-Pro®"-water dispersion and begin heating; while stirring continuously heat this blend to 70-73 C. Weigh 1-8 into another container and begin stirring and heating. Heat the fatty acid-containing blend to 70-73 C. Add 9. Add the "Oat-Pro®"-containing blend which should be at 70-73 C to the fatty acid-containing blend which should also be at 70-73 C. Begin cooling while continuing to stir. Cool the batch to 60 C and add 10. Cool to 25-30 C and add the remainder of the ingredients.

Pressurization Procedure:

To the above formula add these:
Propellants 114:112 at a ratio of 42 : 58

1. Above concentrate	88 g
2. Propellant blend given above	12 g

No. 2

(Proteinized Aerosol)

Stearic Acid (triple pressed)	4.50
"Crodacid" Myristic Acid	1.50
Sorbitol Solution (70%)	10.00
Triethanolamine	4.00
"Crodyne" BY 19	2.00
Water	71.00
Preservative, Perfume	q.s.
Propellants 12/114 (40 : 60)	7.00

Heat oil phase to 75-80 C; water phase to 70. Add water to oils with efficient stirring but do not aerate. When uniform, cool to 45 C. Add desired perfume and preservative, fill off and pressurize cans.

No. 3

(Aerosol)

Concentrate:	
Stearic Acid	6.80
Triethanolamine	3.70
"Emcol" CD-17	0.50
"Witcamide" 5170	0.50
"Emcol" L32-45	0.25
"Emcol" RHP	1.00
Glycerin	2.00
Water	84.95
Perfume	0.30
Aerosol:	
Concentrate	96.50
Propellant A-46	3.50

Mix all ingredients, except perfume, and heat to 70 C while stirring. Cool to 40 C and add perfume. Cool to room temperature before pressurizing.

No. 4

(Gel)

"Witcamide" 5170	20.250
Postassium Laurate (40%)	5.625
Water	63.625
Propylene Glycol	5.000
Cetyl Alcohol	0.500
Pentane	4.000
Isobutane	1.000

Mix the first five ingredients together and heat until clear. Pour into aerosol containers, add hydrocarbons and pressurize with 25 psi nitrogen. Shake contents until clear.

This formulation emerges from the container a clear gel. As the gel is rubbed on the skin, it expands and becomes a rich, creamy lather.

No. 5

(Cream)

Stearic Acid (triple pressed)	3.80
"Neofat" 255	0.90
Mineral Oil 65/75	0.50
Potassium Hydroxide (20% aqueous sol'n.)	0.90
Triethanolamine (85%)	1.68
Glycerin	2.70

No. 6

(Women's)

A	"Veegum"	3.0
	Water	83.5
B	Glycerin	2.0
	Sorbitol (70% sol'n)	3.0
	"Triton" X-100	3.0
C	"Myrj" 45	5.0
D	"Vancide" 89RE	0.5

Add the "Veegum" to the water slowly, agitating continually until smooth. Add B to A and heat to 70 C. Heat C to 75 C, add to A and B and mix. Cool with agitation and add D at approximately 50 C. Continue mixing until cool.

Brushless Shaving Cream

FORMULA NO. 1

A	Stearic Acid (triple pressed)	8.00
	"Lexate" IL	4.00
	"Lexemul" 561	3.00
	Stearyl Alcohol	1.00
B	Glycerin	10.00
	Triethanolamine (85%)	0.75

"Lexgard" M	0.15
"Lexgard" P	0.05
Water	up to 73.05
Perfume	q.s.

Weigh and melt the ingredients of A, stir until homogeneous and heat mixture to 70-75 C. Charge the ingredients of B into a separate vessel equipped with an agitator and provisions for heating and cooling. Dissolve the "Lexgard"s with heating and agitation and bring temperature of completed B to 70-75 C. Gradually add A to B with vigorous agitation and when addition is complete, reduce agitation and cool to 40-45 C. Add and disperse perfume as required; cool to 35 C and package. Consistency develops fully after 24 h at room temperature.

NO. 2

1.	"Neo Fat" 265	6.00
2.	"Emersol" 233	6.00
3.	"Neo Fat" 1654	13.50
4.	"Neo Fat" 1853	7.50
5.	"Emersol" 132	4.00
6.	Propyl Paraben	0.10
7.	BHA	0.10
8.	Methyl Paraben	0.10
9.	Deionized Water	q.s. 100.00
10.	Propylene Glycol	15.00
11.	"Oat-Pro®"	5.00
12.	Potassium Hydroxide	19.54
13.	Perfume	q.s.

Weigh 1-7 into a container and begin heating this blend to 70-73 C. Commence agitation using a mixer which provides good stirring of heavy creams. Weigh 9 and add 11 using a mixer equipped with a stirrer capable of imparting relatively high shear stress (an agitator equipped with a propeller type stirrer is recommended). Stir until the "Oat-Pro®" is completely dispersed and hydrated and no lumps of "Oat-Pro®" are visible. Add the remainder of the individual ingredients to the water with the exception of 13 and adjust the temperature to 70-73 C. When the water-containing blend and fatty acid blend are both at 70-73 C add the water blend to the fatty acids. Continue stirring and commence cooling. Cool the batch to 30–35 C; add perfume. Permit to remain overnight; resume stirring for one hour the following day and package.

No. 3

Stearic Acid	2.00
Mineral Oil	10.00
"Ammonyx" SO	10.00
D & C Silicone 05-0158A	5.00
BTC 2125M	0.10
"Onyxide" 500	0.05
Deionized Water	72.85

Lotions

FORMULA No. 1

(Astringent)

Zinc Oxide	10.0
Talc	10.0
Lanolin	10.0
Peanut Oil	40.0
"Tween" 80	2.0
Water	28.0
Zinc or Aluminum Phenolsulfonate	3–10%

Heat the lanolin, peanut oil and "Tween" 80 to 60 C. Mix well and cool to room temperature. Add the zinc oxide and talc, milling if necessary. Add the phenolsulfonate salts dissolved in water.

No. 2

(Astringent)

Alcohol	13.5
Propylene Glycol	10.0
"Tween" 20	3.0
Boric Acid	4.0

Zinc or Aluminum Phenolsulfonate	1.0
Perfume	0.5
Water	68.0

Mix propylene glycol and "Tween" 20 with water. Dissolve boric acid and phenolsulfonate salt in the aqueous solution.

No. 3

1.	Deionized Water		60.00
2.	"Laponite" XLG		2.00
3.	"Oat-Pro®"		3.00
4.	Chlorhydrol (50%)		3.00
5.	Ethyl Alcohol (SDA-40)		16.50
6.	Methyl Paraben		0.10
7.	Deionized Water	q.s.	100.00
8.	Perfume		q.s.

Weigh 1 and add 2 while stirring so to impart high shear stress. After 2 has completely dispersed and hydrated add 3, 4, 5, 6, 7, and 8.

Freckle and Dark Skin Remover

Fretinoin	0.1
Hydroquinone	5.0
Dexamethasone	0.1
Hydrophilic Ointment (USP)	94.8

Apply twice daily.

Skin Bleaches

FORMULA NO. 1

Cetyl Alcohol	10.00
"Lexemul" 561	10.00
"Pluronic" L64	0.25

H_2O_2 (35%)	17.14
Water	up to 62.61
H_3PO_4 (85%) to pH 3.5–4.0	0.01 to 0.10%

Charge water into making tank and heat to 70-75 C. Weigh cetyl alcohol, "Lexemul" 561 and "Pluronic" L64 into separate container; melt ingredients, and mix until homogeneous. Bring temperature of melted wax to 70-75 and add to the heated water phase with vigorous agitation. Reduce agitation and cool to 30 C. Add H_2O_2 (35%) at 25-30 C and stir until uniformly incorporated. Adjust pH with H_3PO_4 as required. Package. Consistency develops fully after 24 h at room temperature.

No. 2

A	"Veegum" K	2.00
	Water	68.05
B	Glycerin	4.00
	Triethanolamine	1.00
C	Stearic Acid	2.00
	Cetyl Alcohol	4.00
	Isopropyl Myristate	2.00
	"Amerchol" L-101	10.00
	"Arlacel" 165	3.00
D	Citric Acid	0.30
	Hydroquinone	3.00
	Sodium Sulfite	0.40
	Sodium Metabisulfite	0.25
	Preservative	q.s.

Add the "Veegum" to the water slowly, agitating continually until smooth. Add B to A and heat to 75 C. Heat C ingredients to 70 C and add to B and A with stirring. While stirring, cool to 65 C, add citric acid. Cool to 55 C, add the hydroquinone. Cool to 45 C, add the sodium sulfite, and then the sodium metabisulfite. Stir until cool and uniform.

NO. 3

A	"Lexemul" AS	12.00
	Cetyl Alcohol	5.00
	"Lexgard" P	0.10
B	Hydroquinone	3.00
	Sodium Bisulfite	0.10
	"Lexgard" M	0.20
	Deionized Water	79.60

Heat A and B to 85 C separately. Add A to B and stir to 35 C and fill jars.

NO. 4

(Baby)

Light Mineral Oil	35.00
White Petrolatum	4.20
"Super Hartolan"	1.25
Ceto/Stearyl Alcohol BP	0.50
"Crodesta" F-110	3.00
Glycerin	1.00
Perfume, Preservatives	q.s.
Water	55.05

Blend oil phase by stirring and heating to 75 C. Add water to oil under high speed mixer. Stir until cool. Fill off.

NO. 5

(Clear)

"Crodafos" N3 Acid	2.5
Triethanolamine	1.0
"Novol"	7.5
"Volpo" 3	2.5
Ethanol SDA-40 (95%)	42.5
Water	44.0

Mix the alcohol and oil phase at room temperature followed by the addition of the water.

No. 6

"Manucol" DH	1.3
Triethanolamine	0.3
Preservative	as required
Glycerol Monostearate	3.0
Stearic Acid	1.0
Water	to 100

Dissolve the "Manucol" DH in the water and then add the other ingredients. Heat the mixture slowly until all the components have melted. Allow the mixture to cool slowly with continuous stirring. Allow the mixture to stand overnight, stir and then pack.

No. 7

(All Purpose)

"Skliro"	2.00
"Polawax" (ceteareth-5)	1.00
Lanolin Alcohols	0.25
White Mineral Oil	2.00
"Emcol" E-607S	1.00
Water	83.55
Glycerin	10.00
"Emcol" E-607L	0.20
Color, Perfume, Preservative	q.s.

Heat oil and water phases separately to 75–80 C. Add water to oil phase with agitation. Maintain agitation while cooling to room temperature; add fragrance at 30–40 C.

		No. 8	No. 9	No. 10
A	Oil Phase			
	Mineral Oil 65/75	23.0	21.0	6.0
	"Pationic" CSL	3.2	5.4	5.4
	"Pationic" ISL	0.8	0.6	0.6
B	Water Phase			
	Glycerin	3.0	3.0	3.0
	Sodium Lactate (60%)	1.0	1.0	—
	Water	69.0	69.0	85.0
	Preservative	q.s.		

Combine A and heat to 80 C. Combine B and heat to 80-85 C. Add B to A while stirring and continue to stir until room temperature is reached.

No. 11

A	"Ceraphyl" 424	2.0
	"Ceraphyl" 375	4.0
	"Cerasynt" SD	3.0
	"Myrj" 52	1.0
	"Promulgen " D	1.0
	Cetyl Alcohol	0.5
B	Deionized Water	58.4
	"Cellosize" QP 4400 (2% aq.)	25.0
	"Ceraphyl" 60	2.0
	Propylene Glycol	3.0
	"BTC" 2125M	0.1

Heat A and B to 80 C. Add A slowly to B at 80 C with continuous agitation. Cool with stirring (avoid aeration) to 25-28 C.

No. 12

(Proteinized)

A	Stearic Acid (triple pressed)	3.00
	Mineral Oil (light)	5.00
	"Lexol" PG 8-10	2.00
	"Lexgard" P	0.10
B	"Lexein" X250	5.00
	Triethanolamine	0.30
	"Lexgard" M	0.15
	"Bronopol"	0.05
	Propylene Glycol	5.00
	Perfume	0.25
	Ethyl Alcohol	3.00
	Water	76.15

Heat A and B separately to 65–70 C until dissolution. Add A to B at 65–70 C with agitation. Cool batch to 45 C and add balance of ingredients. Cool batch to 30 C and fill.

No. 13

(Silicone)

A	Silicone Fluid (350 centistokes)	10.00
	"Brij" 52 (polyoxyethylene-2-cetyl ether)	3.95
	"Arlasolve" 200	3.55
	Triethanolamine	0.20
B	Water	81.80
	"Carbopol" 934	0.20
	Methyl para Hydroxybenzoate	0.18
	Propyl para Hydroxybenzoate	0.02
C	Perfume	0.10

Disperse the "Carbopol" 934 and para hydroxybenzoates in the water with rapid agitation. Heat B to 65 C and A to 60 C. Add B to A with rapid agitation. Add C at 40 C. Add water to compensate for loss during preparation. Adjust pH to 6.5–7.0 with triethanolamine if necessary.

No. 14

(Silicone)

Oleth-3 Phosphate	2.2
Cetyl Alcohol	.55
Lanolin (USP)	.55
Isopropyl Myristate	2.2
Silicone Fluid SF-96	2.2
Stearic Acid	3.3
Triethanolamine	.55
Propylene Glycol	5.5
Distilled Water	82.18
Cellulose Gum, CMC-7HF	.55
Methyl Paraben	.22
Perfume	as needed

Combine all oil-phase ingredients and bring to 80 C. Bring the triethanolamine, propylene glycol, and water mixture to 80 C and add to the oils with good mechanical agitation. Cool to 60 C and add the cellulose gum/preservative solution with slow agitation. Cool to 35 C with slow agitation, blend in the perfume, and fill.

No. 15

(Cooling)

A	"Amerlate" P	1.0
	"Arlasolve" 200	3.0
	"Arlamol" E	5.0
B	"Natrosol" 250 HR	0.2
	"Carbopol" 934	0.5
	Deionized Water	59.8
C	Triethanolamine	0.5
D	Alcohol SDA-40	30.0

Heat A to 75 C. Disperse "Natrosol" in half of the water and "Carbopol" in the other half; combine. Heat B to 75 C. Add B to A with agitation. Stir five minutes and add C. Cool to 35 C and add D. Stir to room temperature.

Skin Freshener

FORMULA NO. 1

1.	"Oat-Pro®" (2% dispersion in deionized water)		50.00
2.	"Carbopol" 940 (2% dispersion in deionized water)		7.50
3.	Triethanolamine		0.20
4.	Allantoin		0.10
5.	Potassium Sorbate		0.10
6.	Methyl Paraben		0.10
7.	Propylene Glycol		3.00
8.	Deionized Water	q.s.	100.00
9.	Perfume		q.s.
10.	Color		q.s.
11.	Alcohol SDA-40		25.00

Heat 1 to approximately 63 C while stirring continuously, add 4, 6, 7, 8, and 2. Continue stirring and cool the batch to approximately 40 C and add 5, 9, 10, and 11.

NO. 2

1.	"Carbopol" 940	0.20
2.	Deionized Water	50.00
3.	Ethyl Alcohol SDA-40	15.00
4.	"Oat-Pro®"	3.00
5.	Triethanolamine	0.20
6.	Methyl Paraben	0.10
7.	Allantoin	0.10
8.	Perfume	q.s.
9.	Color	q.s.
10.	"Dowicil" 200	0.10

Weigh 2; commence stirring. Slowly add 1. When the "Carbopol" has completely hydrated proceed to weigh 3 and add 6 and 7. Stir until these items dissolve and add this to the "Carbopol" water dispersion. Add the remainder of the formulation ingredients while stirring continuously. Stir after the batch has been permitted to age for 12–15 h and fill.

No. 3

Alcohol	28.4
Propylene Glycol	6.7
Boric Acid	2.0
Menthol	0.1
Perfume	0.7
Water	58.9
Zinc or Aluminum Phenolsulfonate	1.1

Moisturizing Cream

FORMULA NO. 1

1.	Stearic Acid (triple pressed)		3.00
2.	Stearyl Alcohol		2.00
3.	Mineral Oil (65/75 saybolt)		7.00
4.	Propyl Paraben		0.10
5.	Methyl Paraben		0.10
6.	Deionized Water	q.s.	100.00
7.	"Oat-Pro®"		2.00
8.	Propylene Glycol		5.00
9.	Triethanolamine		1.00
10.	Potassium Sorbate		0.10
11.	Perfume		q.s.
12.	Color		q.s.

Weigh 1, 2, 3, and 4 into a container; commence stirring and heat to 70 C. In another container weigh 5, 6, 7, 8, and 9; commence stirring while heating to 70–73 C. Cool to 40 C; add 10, 11, and 12.

No. 2

1.	"Polawax"		2.25
2.	Stearyl Alcohol		2.50
3.	Mineral Oil (65/75 saybolt)		12.00
4.	"Cetiol" V		2.00
5.	Methyl Paraben		0.10
6.	Propyl Paraben		0.10
7.	Deionized Water	q.s.	100.00
8.	"Oat-Pro®"		2.00
9.	Glycerin		5.00
10.	"Carbopol" 940		0.10
11.	Triethanolamine		0.10
12.	Sodium Benzoate		0.10
13.	Color		q.s.
14.	Perfume		q.s.

Weigh ingredients 1-6 and commence stirring while heating to about 72 C. In a separate container add 10 to 7 while mixing continuously then weigh and add 8 and 9. Heat the water phase while stirring continuously to about 72 C and add to the oil phase which is also at this temperature. Add 11, continue mixing and cool the batch to about 40 C at which temperature add 12, 13, and 14. Fill at 25-30 C.

No. 3

A	"Neo-Fat" 16	3.35
	Cetyl Alcohol	1.50
	Isopropyl Myristate	3.00
	Hexachlorophene	0.05
	Methyl para Hydroxybenzoate	0.10
	Propyl para Hydroxybenzoate	0.10
B	Perfume	q.s.
C	Triethanolamine	0.90
	Propylene Glycol	3.00
	Deionized Water	q.s.

Heat A to 70 C. Heat C to 75 C. Add C to A with agitation and cool to 40 C. Add B, mixing thoroughly.

No. 4

(Acidic)

A	"Cerasynt" 945	12.0
	"Ceraphyl" 375	5.0
	Mineral Oil	10.0
	"Promulgen" D	2.5
	Cetyl Alcohol	1.0
B	Deionized Water	59.0
	"Ceraphyl" 60	3.0
	Propylene Glycol	5.0
C	"Dowicil" 200 (10% aq.)	2.5

Heat A and B separately to 80 C. Add slowly A to B with constant agitation at 80 C. Cool to 40 C and add C. Continue cooling to 25-28 C.

No. 5

A	"Veegum"	5
	Water	75
B	Propylene Glycol	2
	"Modulan"	2
	Petrolatum (white)	10
	"Amerchol" L	6
	Preservative	q.s.

Add the "Veegum" to the water slowly, agitating continually until smooth. Add B to A and heat to 70 C. Mix until uniform.

No. 6

A	"Carbopol" 941	0.5
	Water	65.5

B	"Amerlate" P	2.5
	"Emulan"	5.5
	"Promulgen" G	3.5
	"Ceraphyl" 424	1.5
C	Triethanolamine	0.5
	Ethanol (95%)	19.5
	Perfume	1.0

Add B to A at 75 C and stir-cool to 45 C before adding C. Then continue stirring down to 30–35 C before taking off.

Note: As C is incorporated, the batch gets very heavy (high viscosity). Propeller mixing is not appropriate at this stage. Side-sweep agitation will be required.

No. 7

(Antiperspirant)

Isopropyl Alcohol	20.0
Propylene Glycol	20.0
"Tween" 20	10.0
Water	50.0
Zinc or Aluminum Phenolsulfonate	3–10%

Dissolve phenolsulfonate salt in 30 ml. of water. Mix all ingredients and stir until clear.

	No. 8	No. 9	No. 10
"Volpo" 3	1.0	—	—
"Volpo" 10	—	3.0	1.5
"Polychol" 5	—	—	1.5
"Crodafos" No. 10 (neutral)	1.0	—	—
"Crodamol" IPM	—	5.0	5.0
Perfume	2.0	1.0	1.0
Ethanol SDA-40 (95%)	40.0	33.0	33.0
"Carbopol" 941	0.4	0.3	0.3
Triethanolamine	0.4	0.3	0.3
Water	55.1	57.4	57.4

Dissolve the emulsifiers and amine and oil in the alcohol. Disperse the "Carbopol" in water until completely dispersed. (Note: it may be desirable to make a 2% stock solution of "Carbopol.") Mix the two phases together.

No. 11

(Acidic)

"Witconol" MST	4.0
Mineral Oil	5.0
Lanolin (USP)	0.5
Cetyl Alcohol	1.0
"Witconol" H-35A	0.5
"Emcol" E-607S	0.5
Silicone (250 cs)	0.4
Water	82.1
Propylene Glycol	6.0
Lactic Acid	to pH 4.5-5.0
Perfume, Preservative	q.s.

Heat oil and water separately to 75-80 C. Add oil to water phase with moderate agitation. Maintain agitation while cooling to below 30 C. Add fragrance at 40 C.

Cosmetic Stick, Emollient

A	"Veegum"	1
	Water	19
B	"Amerchol" L-101	15
	Isopropyl Myristate	24
	Isopropyl Esters of Lanolin Fatty Acids	10
	Stearic Acid	8
	"Microcrystalline Wax"	23
	Preservative	q.s.

Add the "Veegum" to the water slowly, agitating continually until smooth. Heat to 75 C. Heat the ingredients in B to 70 C with stirring. Add A to B with stirring and stir until smooth. Pour into mold and allow to harden at room temperature (ca. 72 F).

Night Cream

FORMULA NO. 1

"Volpo" 3	2.0
Light Mineral Oil	14.0
White Beeswax	9.0
Paraffin Wax (140 F)	5.0
"Crodamol" CSP	1.0
Lanolin USP	7.9
Sesame Oil	16.0
BHA (butylated hydroxy anisole)	0.1
Borax	0.9
Perfume, Preservatives	q.s.
Water	44.1

Add the water phase to the oils at 70 C with mechanical agitation. Stir and force cool to 40-50 C. Perfume and continue to cool to as low a temperature as possible before cream solidification. Homogenize at room temperature for better stability.

NO. 2

1.	"Arlacel" 165		5.00
2.	Cetyl Alcohol		10.00
3.	Mineral Oil (65/75 saybolt)		25.00
4.	Propyl Paraben		0.10
5.	Silicone Fluid (200/350 centistokes)		0.50
6.	Glycerin		5.00
7.	Methyl Paraben		0.10
8.	"Oat-Pro®"		1.50
9.	Deionized Water	q.s.	100.00

10.	Potassium Sorbate	0.10
11.	Perfume	q.s.
12.	Color	q.s.

Heat the oil phase (1-5) to 70-73 C while stirring. Heat the aqueous phase (6-9) to 70-73 C; while stirring add the aqueous to the oil phase; cool to about 40 C and add 10, 11, and 12. Fill at 25-30 C.

No. 3

"Polysynlane"		15.0
Paraffin Wax		2.0
Lanolin Oil		4.0
Hydrogenated Lanolin		6.0
Beeswax		3.0
Stearic Acid		1.5
Glyceryl Monostearate		2.5
Isopropyl Myristate		6.0
PEG-200 Monostearate		2.0
Potassium Hydroxide		0.2
Propylene Glycol		6.0
Preservatives and Perfume		q.s.
Water	q.s.	100.0

Cold Cream

Formula No. 1

"Polysynlane"	32.0
Mineral Oil	4.0
Paraffin Wax	4.0
Isopropyl Myristate	8.0
Beeswax	3.0
Lanolin	8.0
Propylene Glycol	4.0
Potassium Hydroxide	0.3
"Arlacel" 40	2.5

P.O.E. Sorbitol		1.0
Stearic Acid		1.5
Perfume and Preservatives		q.s.
Water	q.s.	100.0

No. 2

(Liquid)

A	"Veegum"	1.0
	Water	69.0
B	Beeswax	1.5
	Spermaceti	1.5
	Mineral Oil	20.0
	Sorbitan Monopalmitate	3.5
	Polysorbate 60	3.5
	Preservative	q.s.

Add the "Veegum" to the water slowly, agitating continually until smooth. Heat to 75 C. Heat B to 80 C and add to A. Mix until cool.

No. 3

(Emollient)

A	"Arlamol" E	20.0
	"Arlasolve" 200	0.8
	"Brij" 72	7.2
	Stearyl Alcohol USP	2.0
B	"Carbopol" 934	0.2
	"Dowicil" 200	0.1
	Deionized Water	69.5
C	Triethanolamine	0.2
D	Perfume	q.s.

Disperse the "Carbopol" in the water. Heat A to 60 C and B to 65 C. Add B to A with good agitation. Add C. Add D between 35–40 C. Pour about 35 C.

No. 4

(Body)

A	Cetyl Alcohol	1.9
	Stearyl Alcohol	3.0
	Isopropyl Myristate	1.3
	Light Silicone Oil	0.8
	Sodium Stearoyl-2-Lactylate	1.1
	Methyl Propyl Paraben	q.s.
B	Water	88.9
	Propylene Glycol	3.0

No. 5

(Body)

A	Cetearyl Alcohol	5.0
	Isopropyl Myristate	2.0
	"Pationic" 145A	2.0
	"Pationic" ISL	2.0
B	Deionized Water	83.6
	Propylene Glycol	5.0
	Sodium Citrate	0.2
	Methyl Paraben	0.2
C	Perfume	q.s.

Combine A, heat to 80 C. Combine B, heat to 82 C. Add B to A slowly with thorough stirring. Continue stirring until 40 C and add C. Finish mixing and package.

No. 6

(Day)

"Polysynlane"	15.0
Stearic Acid	3.0
Cetanol	1.5
"Arlacel" 60	2.0

"Tween" 60		1.0
Propylene Glycol		6.0
Perfume and Preservatives		q.s.
Water	q.s.	100.0

No. 7

(Hand)

A	"Cerasynt" Q	13.0
	"Ceraphyl" 424	1.0
	"Ceraphyl" 375	4.0
	Sperm Wax	4.0
	"Preservatol"	0.1
B	Deionized Water	57.7
	Glycerin	10.0
C	Deionized Water	10.0
	Allantoin	0.2

Heat A and B separately to 80 C. Add A to B at 80 C. Cool to 60 C with constant agitation and add C (which is heated to 60 C). Continue cooling to 25–28 C.

No. 8

(Cleansing)

1.	Beeswax (USP)		14.00
2.	Lanolin		10.00
3.	Mineral Oil (65/75 saybolt)		40.00
4.	Glyceryl Monostearate (self-emulsifying)		1.25
5.	"Polawax"		0.50
6.	Deionized Water	q.s.	100.00
7.	Borax		1.30
8.	Methyl Paraben		0.10
9.	Perfume		q.s.
10.	Color		q.s.
11.	"Oat-Pro®"		1.50

Weigh ingredients 1, 2, 3, 4, and 5 into a container and heat while stirring to approximately 72 C. Weigh ingredients 6, 7, 8 into a container and heat to 72 C while stirring. At 72 C add 11 and add the water phase of the emulsion to the oil phase while stirring. Cool to about 40 C; perfume and color. Fill at 25-30 C.

No. 9

(All-Purpose)

A	"Lexemul" 561	5.00
	Cetyl Alcohol	5.00
	"Lexol" 3975	9.00
	Mineral Oil 125/135	10.00
	Stearic Acid (triple pressed)	3.00
B	Water	up to 64.30
	Propylene Glycol	2.50
	Triethanolamine	1.00
	"Lexgard" M	0.15
	"Lexgard" P	0.05
	Perfume	q.s.

Charge water into making tank and heat to 65-70 C. Begin agitation and add the balance of ingredients of B, agitating until dissolved. Weigh and melt the ingredients of A in a separate vessel, stir until homogeneous and bring the temperature to 65-70 C. Add A to B with vigorous agitation, cool with stirring to 40-45 C and stir in perfume. Cool to 30-35 C and package. Product consistency develops fully after 24 h at room temperature.

No. 10

(Glycerin)

"Manucol" KMR	2.7
Glycerin	6.0
Methyl para Hydroxybenzoate (preservative)	0.1
Perfume and Color	as required
Water	to 100

Disperse the perfume and color in the water. Disperse the "Manucol" KMR into the glycerin and preservative. Add the "Manucol" KMR dispersion to the water with high shear mixing. Continue stirring until dissolved.

No. 11

(Protective Hand Cream)

Stearic Acid	15.0
"Span" 60	2.0
"Tween" 60	1.5
Zinc Stearate	5.0
Glycerin	6.0
CMC-7HF (2% sol'n)	37.5
Water	33.0
Preservative	q.s.

Melt all fats and zinc stearate and heat to 90 C. Dissolve glycerin and preservatives in water and heat to 90 C. Add the second batch to the first batch with rapid agitation. Drop temperature to 55 C and add slowly to the CMC solution.

No. 12

(Powdered Cream)

A	Mineral Oil (70 cps)	3.00
	Mink Oil	1.40
	Isostearyl Isostearate	1.00
	Isostearyl Alcohol	1.00
	Isopropyl Lanolate	0.40
	Diethanolamine	0.20
	Stearic Acid XX (double pressed)	0.80
	"Arlacel" 165	0.50
	Propylene Glycol	0.50
	"Preservatol"	0.04
	Water	1.16
B	"Avicel" RC-591 MCC	90.00

Heat all Phase A ingredients together to 75-80 C, hold for 10-15 min to saponify, then cool 50 C. Add A to B slowly with continuous agitation until completely dispersed. Tray dry at 45-50 C to evaporate water content. Add water to make a cream.

No. 13

(Vitamin E)

Paraffin Wax		7
Petrolatum		42.5
Isopropyl Myristate		4
Cetyl Alcohol		2
Anhydrous Lanolin		4
Sorbitan Monooleate		4
Magnesium Sulfate		0.2
"Sorbo"		3
Vitamin E		0.4
Preservative		0.2
Water	q.s.	100

No. 14

(Vitamins A, E)

Stearic Acid	13
"Lantrol"	5
Glyceryl Monostearate (S.E.)	8
"Tween" 85	2
"Span" 85	1
Peach Kernel Oil	5
Vitamin A Palmitate	0.1
Vitamin E	0.02
Triethanolamine	1
Methyl Paraben	0.2
Water	52.2

NO. 15

(Rolling Massage)

A	"Veegum"	2
	Water	79
B	Triethanolamine	1
C	Stearic Acid	3
	Paraffin Wax	15
	Perfume and Preservative	q.s.

Add the "Veegum" to the water slowly, agitating continually until smooth. Add B and heat to 65-70 C. Heat C to 65-70 C. Add A and B to C and mix until the temperature drops to 50 C. Allow to stand overnight. Remix before packaging.

Vanishing Cream

FORMULA NO. 1

A	"Lexemul" 530	4.00
	Stearic Acid (triple pressed)	16.00
	Mineral Oil (light)	2.00
	"Lexgard" P	0.10
B	Water	74.20
	Glycerin	3.00
	Triethanolamine	0.50
	"Lexgard" M	0.20

Heat A and B to 65 C. Add A to B and with continuous mixing cool to 35 C. Fill.

NO. 2

Stearic Acid	15.0
Cetanol	1.5
Glyceryl Monostearate N.S.E.	1.5

"Polysynlane"		7.0
Potassium Hydroxide		0.5
Glycerin		5.0
Perfume and Preservatives		q.s.
Water	q.s.	100.0

	No. 3	No. 4
"Codesta" S20	–	3.0
"Crodoafos" N.3 Acid	4.0	4.0
Lanolin (USP)	2.0	–
"Novol"	1.5	1.5
Apricot Kernel Oil (N.F.)	2.0	2.0
"Crodamol" CSP	1.0	–
Stearic Acid (triple pressed)	12.0	12.0
Polyethylene Glycol 4000	10.0	10.0
Triethanolamine	2.6	2.6
Perfume, Preservatives, and Color	q.s.	q.s.
Water	64.9	64.9

Melt the oils and waxes and bring to 75 C. Heat the aqueous phase to the same temperature or slightly higher and when both phases are dissolved add the water to the oils with mechanical agitation. Hot fill at 50 C after perfuming.

Cleansing Cream

FORMULA NO. 1

"Amerchol" CAB	5.0
"Amerlate" P	2.0
Polysynlane	30.0
Beeswax	10.0
"Arlacel" 60	2.0
Ozokerite	5.0
"Carbopol" 940	0.2
Triethanolamine (10% sol'n)	2.0
"Tween" 60	3.0

Propylene Glycol		4.0
Preservatives and Perfume		q.s.
Water	q.s.	100.0

No. 2

1.	Beeswax (USP)		14.00
2.	Lanolin		10.00
3.	Mineral Oil (65/75 saybolt)		40.00
4.	Glyceryl Monostearate (self-emulsifying)		1.25
5.	"Polawax"		0.50
6.	Deionized Water	q.s.	100.00
7.	Borax		1.30
8.	Methyl Paraben		0.10
9.	Perfume		q.s.
10.	Color		q.s.
11.	"Oat-Pro®"		1.50

Weigh ingredients 1, 2, 3, 4, and 5 into a container and heat while stirring to approximately 72 C. Weigh ingredients 6, 7, 8 into a container and heat to 72 C while stirring. At 72 C add 11 and add the water phase of the emulsion to the oil phase while stirring. Cool to about 40 C; perfume and color. Fill at 25-30 C.

No. 3

(Liquid)

1.	"Promulgen"		2.50
2.	Mineral Oil (65/75 saybolt)		20.00
3.	Lanolin		1.00
4.	Isopropyl Myristate		5.00
5.	"Emcol" 607S		0.25
6.	"Carbitol"		0.75
7.	Deionized Water	q.s.	100.00
8.	"Oat-Pro®"		2.00
9.	Glycerin		5.00
10.	Methyl Paraben		0.10

11.	Propyl Paraben	0.10
12.	"Dowicil" 200	0.10
13.	Perfume	q.s.

No. 4

(Friction)

A	"Veegum"	1.75
	Water	56.50
B	Beeswax	1.32
	Spermaceti	1.32
	Light Mineral Oil	17.40
	Sobitan Monopalmitate	3.05
	Polysorbate 60	3.05
	Cetyl Alcohol	2.61
C	"Nytal" 100	13.00
	Preservative	q.s.

Add the "Veegum" to the water slowly, agitating continually until smooth. Heat A to 75 C. Heat B to 80 C. Add B to A and mix thoroughly. Add C and continue mixing until temperature reaches 40 C.

No. 5

(Gel)

"Pluronic" F-127 Polyol	22
"Miranol" 2 MCAS Modified	10
Laneth-10 Acetate	2
Propylene Glycol	2
Water	64
Bactericide, Perfume	q.s.

Dissolve the "Pluronic" F-127 polyol in cold water (5–10 C) by adding flakes slowly to a well-stirred vortex. Mix slowly while maintaining temperature. Add the other ingredients and then warm to room temperature, at which a crystal clear, ringing gel is formed.

No. 6

(Biostatic)

A	"Veegum" K	1
	Water	22
	"Triton" X-202 (30% solids)	66
	Propylene Glycol	5
B	"Vancide" 89RE	2
	"Amerchol" L-101	1
	"Solulan" 16	3

Add the "Veegum" K to the water slowly, agitating continually until smooth. Add the rest of A and heat to 55 C. Disperse the "Vancide" 89RE in the B phase while heating to 60 C. Add B to A and stir until cool.

No. 6

(Scrub)

"Bentonite" 670	10.4
Titanium Dioxide	0.8
"Maprofix" TAS	8.0
Fine Pumice	0.8
"Onamer" M	2.6
Propylene Glycol	4.0
Perfume	0.2
Distilled Water	73.2

No. 7

(Detergent)

A	"Veegum"	1.0
	Water	50.7
B	Cetyl Alcohol	0.3
	Stearyl Alcohol	0.3
	Lanacet	1.0
	"Nimlesterol" D	5.0
	Stearic Acid	4.4

"Vanseal" CS	3.3
"Pluronic" 25R8 Polyol	12.0
Sodium Lauroylisethionate	20.0
Aromox C/12W	2.0
Preservative and Perfume	q.s.

Add the "Veegum" to the water slowly, agitating continually until it is smooth. Heat A to 75 C. Heat B to 70 C with slow mixing until uniform. Add A to B with slow agitation. Allow to cool to 40 C, add preservative and perfume, and package while warm. Product is an opaque, viscous cream.

No. 8

(Bar)

A	"Vancide" 89RE	1.0
	"Veegum" F	1.0
B	"Igepon" AC-78 (83% solids)	57.3
C	Cetyl Alcohol	2.0
	Glyceryl Monostearate A.S.	5.5
	Stearyl Alcohol	7.5
	"Modulan"	3.0
	Polyethylene Glycol 6000	13.0
	Citric Acid	0.7
	Water	9.0

Blend A in part of B, then add the balance of B and blend well. Heat C to 70–75 C. Add A and B to C with agitation until uniform. Press into bar or cake.

The final pH should be about 5.0.

Hand Cleaners

Formula No. 1

"Tergitol" 25-L-9	9.0
"Isopar" M	35.0
Mineral Oil	3.0

Oleic Acid	6.0
Lanolin	1.0
Triethanolamine	2.1
Monoethanolamine	0.43
Propylene Glycol	3.0
Water	q.s.
Dye	q.s.
Perfume	q.s.

No. 2

A	"Veegum"	2.5
	Water	67.0
	Triethanolamine	2.5
	Propylene Glycol	5.0
B	Stearic Acid	4.0
	Cocoyl Sarcosine	6.0
	Cetyl Alcohol	1.0
	Glyceryl Monostearate	3.0
	Mineral Oil	5.0
	"Lantrol"	4.0
	Preservative	q.s.

Add the "Veegum" to the water slowly, agitating continually until smooth. Add balance of A and heat to 75 C. Heat B to 70 C. Add A to B, agitating continually until smooth.

No. 3

(Waterless)

Lanolin	20.0
"Polawax"	8.0
Methyl Paraben USP	0.2
"Isopar" H	40.0
"Carsonol" SLS Special	2.0
Perfume	q.s.
Water	29.8

Blend together all ingredients except water. Warm to 50–55 C. Heat water to same temperature and add to the oils with efficient stirring. Cool to approximately 35 C with stirring and fill.

No. 4

(Waterless)

White Mineral Oil	10.0
Water	43.2
Carboxymethyl Cellulose	1.0
Potash (100%)	1.8
Oleic Acid	9.0
Deodorized Kerosene	35.0
Perfume	as required

Carboxymethyl cellulose is dispersed in water with the aid of heat. Use warm water for the preparation. The resulting gel should be free from lumps. After the gel has cooled add potash (as a 45% solution) with agitation.

The oleic acid, deodorized kerosene and white mineral oil are mixed together in a separate container and the blend is then added slowly to the potash-carboxymethyl cellulose-water mixture with slow agitation. High speed agitation should be avoided.

The consistency of the above emulsion can be varied, if desired, by either increasing or decreasing the amount of carboxymethyl cellulose.

	No. 5	No. 6
"Polawax"	5.0	—
"Miranol" C2M	40.0	—
"Carsonol" SLS	17.0	—
"Carsonon" N9	—	20.0
"Polychol" 5	—	1.0
"Volpo" 3	1.5	—
"Super Hartolan"	0.5	0.5
"Ottasept" Extra PCMX	1.0	1.0
Propylene Glycol	5.0	5.0
White Petrolatum	1.0	1.0
Water	29.0	71.5

NO. 5: Blend the water phase and warm to 60 C. Warm the oil phase until the "Super Hartolan" has melted with the "PCMX" and then add to the water phase slowly with stirring, cool to room temperature with slow agitation.

NO. 6: Warm both phases separately to 55 C and disperse the "Carsonon" N9 well. Add the water phase to the oil phase with stirring and continue to cool to room temperature with slow agitation.

Barrier Skin Cream

"Cosmowax"	10.0
"Novol"	4.0
"Solan"	3.0
Silicone Oil 200	5.0
Mineral Oil	15.0
Water	63.0
Preservatives, Perfume	q.s.

Heat the oils to 70 C, heat the water to 70 C and dissolve preservatives in the water. Combine with mixing and cool down to 50 C before adding the perfume.

Barrier Skin Spray

"Emcol" D70-30C	2.1
Propellants 152A/11 (37 : 63)	97.9

Valve: Vapor-tap valve with standard .016 in. actuator.

Comments: When sprayed on the skin and rubbed, "Emcol" D70-30C provides an invisible protective coating. The skin remains water-repellent in the presence of detergents and household solvents.

Soap

FORMULA NO. 1

(Liquid)

"Lonzol" LS-300 (30% sol'n)	25.00
"Barlox" C (30% sol'n)	2.00
"Unamide®" CDX	3.00
Ethylene Glycol Monostearate	1.00
Sodium Chloride	0.50
Citric Acid	0.12
Water	68.38

pH = approx. 7.0

NO. 2

"Lonzol" LS-300 (30% sol'n)	25.00
"Barlox" C (30% sol'n)	2.00
"Unamide®" CDX	3.00
"Morton" Latex E-295	0.50
Sodium Chloride	0.50
Citric Acid	0.12
Water	68.88

pH = approx. 7.0

NO. 3

A	"Veegum" HS	1.0
	Water	42.0
B	Potassium Hydroxide	2.0
	Water	37.5
	Propylene Glycol	2.5
	Sodium Lauryl Sulfate (30%)	6.0
C	Oleic Acid	9.0

Add "Veegum" HS to water slowly, agitating continually until smooth. Heat to 75 C. Dissolve potassium hydroxide in the water, mix in

B ingredients. Heat to 75 C. Add B to A, mix until uniform. Heat C to 90 C, add to A and B. Mix until smooth.

Soap, Emollient

"Gelamide" 250	1.00
Soap Chips	96.50
"Tinopal"	0.03
Titanium Dioxide	0.20
"Sequestrene" Na4	0.10
Lanolin (anhydrous)	0.50
Pigment	0.07
"Tenox" 2	0.10
Perfume	1.50

Makeup

FORMULA NO. 1

(Base)

A	"Veegum" HV	2.00
	Sodium Carboxymethylcellulose (low visc.)	0.10
	Water	48.45
B	"Darvan" No. 1	0.30
	Propylene Glycol	15.00
	Water	4.40
C	Allantoin	0.50
	Resorcinol	0.50
	Colloidal Sulfur	1.00
	Titanium Dioxide	3.50
	Iron Oxides	0.25
	Kaolin	1.00
	"Nytal" 400	8.00
D	"Carbowax" 400	10.00
	Polyethylene Glycol 400 Monostearate	5.00
	Perfume and Preservative	q.s.

Dry blend the "Veegum" and the CMC. Add to the water slowly, agitating continually until smooth. Add B to A and mix thoroughly. Heat to 70 C. Blend C and grind in D. Heat to 65 C. Add C and D to A and B. Mix until cool.

No. 2

A	"Veegum"	0.5
	Sodium Carboxymethylcellulose (low visc.)	0.5
	Water	59.3
B	"Darvan" No. 1	0.3
	Triethanolamine	1.0
	Propylene Glycol	5.0
	Water	8.4
C	"Nytal" 400	6.0
	Kaolin	0.8
	Titanium Dioxide	3.0
	Iron Oxides	0.2
D	Stearic Acid	2.0
	Propylene Glycol Monostearate	0.5
	Lanolin	3.5
	Mineral Oil	4.0
	Isopropyl Myristate	5.0
	Preservative	q.s.

Dry blend the "Veegum" and CMC and add to the water slowly, agitating continually until smooth. Add A to B. Micropulverize C, then add to A and B. Heat to 60–65 C. Heat D to 70 C and add to other components. Mix until cool.

No. 3

(Matte)

A	"Veegum"	2.60
	Sodium Carboxymethylcellulose (low visc.)	0.40
	Water	42.40

B	"Darvan" No. 1	0.30
	Propylene Glycol	5.00
	Water	12.30
C	"Nytal®" 400	18.50
	Kaolin	1.30
	Titanium Dioxide	3.70
	Iron Oxides	1.50
D	Isopropyl Myristate	5.00
	Sorbitan Monolaurate	0.75
	Polyoxyethylene Sorbitan Monolaurate	2.25
	Stearyl Alcohol	2.00
	"Amerchol" L-101	2.00
	Perfume and Preservative	q.s.

Dry blend the "Veegum" and the CMC. Add to the water slowly, continually agitating until smooth. Micropulverize C. Add to B and mull to a smooth paste. Add the paste to A and heat to 60-65 C. Heat D to 70 C and then add to other components. Mix until cool.

No. 4

(Stick)

Calcium Carbonate	55.0
Talc	26.50
Titanium Dioxide	8.0
Corn Starch	2.25
Kaolin	5.50
Zinc Stearate	2.50
Para Hydroxybenzoate Ester	0.25
Yellow Ochre	1.0
Red Iron Oxide	0.75
Brown Iron Oxide	0.10

65 parts of the mixture is added to 35 parts of a 1.4% aqueous solution of "Veegum." The binder solution was thoroughly mixed with the dry ingredients to provide a granulated mixture which was subsequently

extruded, dried, and packaged.

No. 5

A	"Schercomid" AME-70	20.0
	"Schercomol" MM	20.0
	"Promulgen" D	5.0
B	Powder Mixture	5.0
C	"Carbowax" 400	10.0
	Micro Crystalline Cellulose	40.0
	Fragrance, Preservative	q.s.

Heat A to 60 C and stir in B. Add 70 C Part C to the mixture of A and B with high speed agitation. Slow agitation to allow entrained air to escape, while stir-cooling to 30–35 C. This gives a very smooth make-up with a "dry feel" due to its nonoily base.

Matte Leg Makeup

FORMULA NO. 1

A	"Veegum"	3.0
	Water	50.2
	"Carboset" 514	16.8
B	Cocoyl Sarcosine	2.0
	Silicone SF-96 (1000)	2.5
	"Carbowax" 400	4.0
	Glycerin	3.0
C	Color	3.5
	"Nytal" 400	6.0
	Magnesium Carbonate	7.0
	Titanium Dioxide	2.0
	Preservative	q.s.

Add the "Veegum" to the water slowly, agitating continually until smooth. Add the "Carboset" and mix until smooth. Combine components in B and then add to A. Micropulverize C and gradually add to A and B with agitation. Mix until uniform.

No. 2

1.	"Amerlate" LFA		3.00
2.	"Industrene" 1288		6.00
3.	"Cetiol" V		13.00
4.	Propyl Paraben		0.10
5.	Talc		11.50
6.	Titanium Dioxide		4.70
7.	Color Pigments		3.30
8.	"Carboset" 514		16.80
9.	Glycerin		5.00
10.	Methyl Paraben		0.10
11.	Deionized Water	q.s.	100.00
12.	Triethanolamine		0.72
13.	Ammonium Hydroxide		0.16
14.	Perfume		q.s.
15.	"Dowicil" 200		0.10
16.	"Oat-Pro®"		3.50

Weigh and add 1, 2, 3, and 4 into a container and begin stirring and heating. Weigh 11 into another container and begin stirring. The use of a variable speed agitator equipped with a propeller type stirrer which is capable of imparting high shearing stress is recommended. Add 8 to 11 and continue stirring; add 16 and stir until the "Oat-Pro®" is completely dispersed and hydrated. Add 9, 10, and 12 and begin heating while stirring to 70–73 C. Continue stirring and add 6 and 7. Begin cooling the batch, cool to 40–43 C and add the remainder of the ingredients. Cool to 25–30 C and package after passing the batch through a colloid mill.

Blemish Masker

Formula No. 1

"Witcamide" 70	18.7
"Witcamide" MAS	8.3
"Witconol" APM	10.0

"Sono-Jell" No. 9	4.0
White Mineral Oil	25.0
Pigment	25.0
Volatile Silicone Oil 7158	9.0

Disperse pigment in mineral oil, volatile silicone oil, "Witconol" APM and "Sono-Jell" No. 9 at 40–50 C using good agitation. Heat to 85–95 C and add "Witcamide" 70 and "Witcamide" MAS. Stir until mixture is uniform. Add fragrance and package at 83–86 C. Hardness or payout can be adjusted by raising or lowering the content of "Witcamide" 70 and "Witcamide" MAS.

No. 2

Castor Oil	65.0
Lanolin	10.0
Isopropyl Myristate	5.0
Beeswax	7.0
Candelilla Wax	7.0
Carnauba Wax	3.0
Ozokerite	3.0

Kaolin and pigments 12–15%.

No. 3

Castorl Oil	29.0
Butyl Stearate	14.0
Petrolatum	5.6
Beeswax	10.5
Paraffin	3.5
Fragrance	q.s.
Ozokerite	7.0
Pigments and Fillers	30.0

	No. 4	No. 5	No. 6	No. 7	No. 8	No. 9	No. 10	No. 11	No. 12	No. 13
Powder Phase:										
Titanium Dioxide	27.2	27.2	27.2	27.2	27.2	27.2	27.2	27.2	27.2	27.2
Kaolin (USP)	5.0	5.0	5.0	5.0	5.0	5.0	5.0	5.0	5.0	5.0
Talc	3.9	3.9	3.9	3.9	3.9	3.9	3.9	3.9	3.9	3.9
"Oxy Rust"	3.1	3.1	3.1	3.1	3.1	3.1	3.1	3.1	3.1	3.1
"Oxy Umber"	0.8	0.8	0.8	0.8	0.8	0.8	0.8	0.8	0.8	0.8
Pigment Wetter:										
Acetylated Lanolin	—	5.0	—	—	—	—	—	—	—	—
Lanolin Alcohol Linoleate	—	—	5.0	—	—	—	—	—	—	—
PPG-5 Lanolin Alcohols	—	—	—	5.0	5.0	—	—	—	—	—
Lanolin Fatty Acids	—	—	—	—	—	5.0	5.0	—	—	—
Hydroxylated Lanolin	—	—	—	—	—	—	—	5.0	5.0	—
Isopropyl Lanolate	—	—	—	—	—	—	—	—	—	5.0
Waxes:										
Ozokerite	4.0	4.0	4.0	4.0	4.0	4.0	4.0	4.0	4.0	4.0
Petrolatum	10.0	10.0	10.0	10.0	10.0	10.0	10.0	10.0	10.0	10.0
Candelilla	3.0	3.0	3.0	3.0	3.0	3.0	2.0	3.0	3.0	3.0
Carnauba	6.0	6.0	6.0	6.0	6.0	6.0	5.0	6.0	6.0	6.0

	No. 4	No. 5	No. 6	No. 7	No. 8	No. 9	No. 10	No. 11	No. 12	No. 13
Liquid Vehicle:										
Mineral Oil (70 visc.)	35.0	30.0	30.0	30.0	20.0	30.0	32.0	30.0	24.0	30.0
Additives:										
Cetyl Alcohol	2.0	2.0	2.0	2.0	2.0	2.0	2.0	2.0	–	2.0
Isopropyl Palmitate	–	–	–	–	10.0	–	–	–	–	–
Myristyl Lactate	–	–	–	–	–	–	–	–	8.0	–
Preservatives, Antioxidants	q.s.	q.s.	q.s.	q.s.	q.s.	q.s.	q.s.	q.s.	q.s.	q.s.

Blend and micronize the powder phase. Heat the remaining ingredients to 85–90 C while stirring slowly. Maintain heat and stirring until free from lumps. Add the powder phase slowly and stir until the dispersion is uniform. Mill or grind. Hold at 85–90 C while stirring slowly to release any entrapped air. Continue stirring while cooling to 70–75 C. Pour into warmed molds and allow to solidify. Molds should be chrome plated and may be silicone treated to aid release. The filled molds may be chilled to expedite unmolding.

Makeup Remover

FORMULA NO. 1

(Water-in-Oil)

A	"Veegum"	2.0
	Water	32.0
B	Mineral Oil N.F. (37 cs)	29.0
	"Hexadecyl" Alcohol	5.0
	"Acetulan"	2.0
	Olive Oil	2.0
	White Petrolatum	4.0
	"PEG" 400	8.0
	Ceresin Wax	1.0
	"Sorbo"	10.0
	"Arlacel" 186	4.0
	"Tween" 80	1.0
	Preservative	q.s.

Add the "Veegum" to the water slowly, agitating continually until smooth. Heat A to 70-75 C. Heat B to 70-75 C. Add A to B with high shear mixing. Mix until cooled to room temperature.

NO. 2

A	Magnesium Aluminum Silicate	1.5
	Water	54.0
	Propylene Glycol	4.0
B	"Pluronic" 25R8 Polyol	8.0
	"Pluronic" 31R4 Polyol	4.0
	Mineral Oil	23.0
	"Trisolan"	2.5
	Acetylated Lanolin	0.5
	Cocoyl Sarcosine	0.5
	Ceteareth-5	2.0
C	Perfume and Preservative	q.s.

Chapter V

DETERGENTS AND DISINFECTANTS

Cold Wool Wash Compound

"Calsoft" T-60	20
"Calsuds" CD-6	20
Optical Brightener	0.1
Preservative	0.1
Water	59.8

Nonphosphate Detergents

No. 1

Sodium Lauryl Sulfate	15.2
"Deselex" PolyBorate	20.0
Sodium Metasilicate	14.3
CMC	0.5
Sodium Sulfate	50.0

No. 2

Nonionic Surfactant	7
"Propasol" Solvent P	5
Sodium Xylene Sulfonate (actives)	4
EDTA, Tetrasodium Salt (actives)	2
Sodium Carbonate	4
Tetrapotassium Pyrophosphate	—
Trisodium Phosphate (anhydrous)	—
Deionized Water	q.s.

Nonphosphate Detergents

FORMULA	No. 3	No. 4	No. 5	No. 6	No. 7	No. 8	No. 9	No. 10	No. 11
Neodol 25-9	10	10	10	15	15	12	10	10	—
LAS	—	—	—	—	—	—	5	5	15
Sodium Carbonate	30	30	30	—	60	30	40	40	40
Sodium Sesquicarbonate	—	10	—	—	—	—	—	—	—
Sodium Citrate · 2H$_2$O	—	—	—	—	—	20	—	—	—
NTA · H$_2$O	—	—	10	—	—	—	—	—	—
Sodium Metasilicate · 5H$_2$O (1/1 SiO$_2$/Na$_2$O)	10	7	10	25	—	—	—	—	—
Sodium Silicate (2/1 SiO$_2$/Na$_2$O ratio)	—	—	—	—	10	10	15	15	15
CMC	1	1	1	2	2	2	3	3	3
Sodium Sulfate	49	42	39	58	13	26	27	27	27

NO. 12

(Low Foam)

"Sodasil" P-598	62.0
"Sulframin" 90 Flakes	2.0
Nonionic	6.0
Carboxymethylcellulose	1.0
Optical Brighteners	0.5
Perfume	0.1
Sodium Sulfate	27.4
"Satintone" No. 1	1.0

	No. 13	No. 14	No. 15
		(High Foam)	
"Sodasil" P-598	52.0	62.0	72.0
"Sulframin®" 90 Flakes	20.0	15.0	10.0
Carboxymethylcellulose	1.0	1.0	1.0
Optical Brighteners	0.5	0.5	0.5
Perfume	0.1	0.1	0.1
Sodium Sulfate	26.4	21.4	16.4

NO. 16

Nonionic Surfactant (100% basis)	8.00
Sodium Carbonate	45.00
Sodium Silicate (100% basis)	—
Sodium Metasilicate Pentahydrate or Equivalent	10.00
Sodium Sulfate	34.50
Brightener System	0.50
CMC (65%)	1.00
"RP"-70	1.00

NO. 17

Sodium Metasilicate (anhydrous)	31.7
Sodium Hydroxide Flakes	31.7
Sodium Carbonate	31.6
"Monaterge" 85	5.0

No. 18

Water	54.7
Tetrapotassium Pyrophosphate (60%)	25.3
Sodium Hydroxide (50%)	10.0
"Monaterge" 85	10.0

Fine Fabric Detergents

FORMULA No. 1

"Maprolyte" LX	25.00
"Onyxol" 201	2.00
"Ammonyx" TDO	2.00
SDA-40 Alcohol	5.00
Deionized Water	66.00

No. 2

"Maprofix" 60N	25.00
"Maprofix" TLS-500	5.00
Urea	5.00
SDA-40 Alcohol	3.00
"Methocel" MC (2% sol'n)	0.20
Deionized Water	61.80

No. 3

"Maprofix" 60S	15.00
"Onyxol" 201	25.00
SDA-40 Alcohol	5.00
Deionized Water	55.00

No. 4

"Maprofix" 60S	30.00
"Onyxol" 201	10.00
"Surco" SXS	10.00
Deionized Water	50.00

No. 5

"Maprofix" 60N	20.00
"Surco" SXS	3.00
"Surco" SF42B	20.00
"Super Amide" GR	7.00
Deionized Water	50.00

Laundry Detergent

FORMULA No. 1

"Bio Soft" HD-100	45.0
Deionized Water	45.1
Triethanolamine	2.0
Ethanol	4.0
"Tinopal" 5 BM Ex Conc.	0.4
Dye, Perfume	3.5

No. 2

	Stabilizer
Water	577.0
"Neodol" 25-9	0.5
"Gantrez" AN-149	9.5
Potassium Hydroxide	18.0
Dyes (and brighteners if desired)	
Carboxymethylcellulose	5.0
Sodium Silicate	40.0
"Neodol" 25-9 or "Neodol" 25-12	100.0
Tetrapotassium Pyrophosphate	250.0

Bring the water temperature to at least 60 C while stirring with moderate agitation. Add the "Neodol" ethoxylate, then slowly add the "Gantrez" AN-149 powder while bringing the mixture temperature to 90 C. Continue stirring with moderate agitation until the mixture becomes clear (about 20 min). At this point the stabilizer reaction is complete. When the solution is clear, add the KOH pellets and dissolve. The stirrer speed is then increased to maximize stirring efficiency through the remainder of the procedure. Very high shear stirring is required, i.e., 650 ft/ min stirrer tip speed.

Dye and brighteners, if desired, are added.

The CMC, sodium silicate, and the bulk of the "Neodol" ethoxylate are added in that order. The mixture is stirred for 5 min. The tetrapotassium pyrophosphate is added, and the final mixture is stirred for 15 min. The finished product can then be bottled and allowed to cool slowly.

With exception of the "Gantrez" AN-149, all other ingredients may be added as water solutions—with appropriate adjustment for the total water content. Care should be taken throughout the procedure to minimize water evaporation.

Dry Laundry Bleach

FORMULA NO. 1

ACL® 85	15-35
Sodium Sulfate	to make 100

If packaged appropriately to exclude moisture and stored away from excessive heat, bleaches of this type are stable for long periods of time. Sodium sulfate is an inert ingredient, used to balance the formulation.

NO. 2

ACL®	15-40
Sodium Tripolyphosphate	30-40
Sodium Silicate and/or Caustic	20-30
Anionic Surfactants	0-3
Optical Brighteners	0-0.3
Carboxymethylcellulose	0-1
Sodium Carbonate, Sodium Sulfate	to make 100

The ingredients of these formulations should be anhydrous to ensure maximum chlorine stability of the product. Since some of the strongly alkaline chemicals, such as the silicates and caustic, are very hygroscopic, moisture should be avoided during manufacture and storage. Formulations of this type tend to have a shorter shelf life than the above.

NO. 3

(Heavy Duty)

	Phosphate	Nonphosphate
"Tergitol"	12.0	12.0
Sodium Tripolyphosphate (anydrous, light density)	25.0	—
Sodium Carbonate (light density)	30.0	40.0
Sodium Silicate	10.0	10.0
Sodium Carboxymethylcellulose	1.5	1.5
"Tinopal" 5 BM (0.63%)	0.7	0.7
Sodium Sulfate, Perfume	q.s.	q.s.

NO. 4

"Tergitol" 25-L-7	30.0
Linear Alkylbenzene Sulfonate (11 type, sodium salt)	5.0
Ethanol (SD-3A, 200°)	6.5
Triethanolamine	10.0
"Tinopal" 5BM Conc.	0.3
Water, Dye, Perfume	q.s.

NO. 5

Sodium Tripolyphosphate (anhydrous)	30.00
LAS or Equivalent (100% basis)	—
LAS or Equivalent (40% basis)	45.00
Sodium Silicate (100% basis)	—
Sodium Metasilicate Pentahydrate	10.00
Sodium Carbonate	10.00
Sodium Sulfate	—
Brightener System	0.50
CMC (65%)	1.00
"RP"-70	1.00

No. 6

"ACL®" 59, "ACL" 60, or "ACL" 66	2-3
Sodium Tripolyphosphate	30-50
Sodium Metasilicate	10-40
Caustic	0-10
Soda Ash or Sodium Sulfate	to make 100

Cloudy Ammonia

Ammonium Hydroxide	10.0
Soap	1.0
Water	89.0

Fabric Softener, Laundry

Formula No. 1

"Accosoft" 550-75	53
Water	47

No. 2

"Accosoft" 620-90	66
Isopropanol	10
Light Mineral Oil	6
Water	23

Use 2/3-1½ oz per 100 lb fabric

No. 3

"Ammonyx SK"	75
Water	25

Use 1-2 oz per 100 lb wash

Laundry Bleach

"CDB"-59	10
Sodium Tripolyphosphate	20-40
Sodium Sulfate	to make 100

Laundry Bluing

Calcocid Blue 3B	1 oz
Water, Warm	1 gal

Laundry Soda

Sodium Bicarbonate	42
Sodium Carbonate	45

Light Duty Detergents

FORMULA NO. 1

"Calsoft" T-60	40
"Calamide" C	2
"Calfoam" ES-30	15
"Pilot" SXS-40	5
Sodium Chloride	2
"Dowicide" A	0.1
Water	30.25

FORMULA	NO. 2	NO. 3
Linear Alkylbenzene Sulfonate (Type 11, sodium salt), Actives	23.0	16.0
Primary Alcohol Ethoxysulfate (ammonium salt), Actives	13.0	4.0

Coconut Diethanolamide (30% minimum		
purity)	5.0	1.0
Sodium Xylene Sulfonate (actives)	3.5	2.0
Ethanol (SD-3A, 200 Pf.)	3.0	—
Water, Dye, etc.	q.s.	q.s.

No. 4

A	"Veegum"	2.0
	Water	89.0
B	"Super Floss"	1.6
	"Snow Floss"	2.4
C	Sodium Hypochlorite (5% sol'n)	5.0

Add the "Veegum" to the water slowly, agitating continually until smooth. Add B to A mixing thoroughly to disperse abrasive particles. Add C slowly agitating until uniform.

Heavy Duty Detergent

FORMULA NO. 1

"Neodol" 25-9	25.0
"Neodol" 25-3S (60% AM)	16.8
Diethanolamine	10.0
Ethanol 3A (added)	7.6
"Tinopal" RBJ 200%	0.025
"Tinopal" 5BM	0.5
Polar Brilliant Blue RAWL 110%	0.005–0.05
"Tergescent" No. 7	0.10
Water	q.s.

No. 2

"Gantrez" AN-149/N25-9 (99/1)	1.0
"Neodol" 25-12	20.0
"Neodol" 25-3S (60% AM)	8.4

KOH	1.0
Sodium Silicate RU (46%)	8.9
K_2CO_3 $1.5H_2O$	25.0
CMC As Is	0.5
"Tinopal" 5BM	0.5
"Tinopal" RBJ 200%	0.025
Polar Brilliant Blue RAWL 110%	0.005-0.05
"Odrene" 22	0.05
Water	q.s.

	No. 3	No. 4
	45.0	40.0
N Sodium Silicate	40.0	40.0
Sodium Hydroxide (50%)	10.0	10.0
"Monaterge" 85	5.0	10.0
Surface Tension		
(dynes/cm—2% sol'n @ 20 C)	30.5	30.3

Use Dilutions:
  Medium Duty — 2-4 oz/gal
  Heavy Duty — 4-8 oz/gal

pH (as is) = 12.7   Foam = Moderate

Charge water, start agitation and then add N sodium silicate. Then add NaOH solution and agitate until clear. Finally add "Monaterge" 85 and continue agitation (approx. 45 min) until solution clears.

No. 5

Water	26,397
"Accodet" HD	14,642
"Tinopal" CBS	82
Dye	q.s.
Perfume	q.s.
Formalin (37%)	82
Bicarbonate of Soda	41

To a mixing tank, charge proper amount of water.

Note: In order to predissolve optical brighteners, baking soda and avoid over-dilution, suggest you retain a portion of water.

With agitation, add proper amount of "Accodet" HD. Mix until homogeneous and clear. With agitation, add proper amounts of optical brightener, dye, perfume and 37% formalin. Mix until homogeneous and clear.

Check pH and solids. If necessary, make adjustments.

pH (as is) — 8.0–9.5
Solids   — 25–30%

pH should be stable over 20 min intervals. **Check pH!** 'When approved, buffer system with 0.1% bicarbonate of soda. **Do not** add buffer until pH is stable and meets specification. The baking soda may be predissolved in a small quantity of water **prior** to it being added to the production mix. Blend until pH is stable, approximately 20 min.

Remove sample for pH and solids. Shut off agitator and allow air to escape from product. When approved, package.

Enzyme Detergent

Alkylbenzene Sulfonate	30.00
Sodium Tripolyphosphate	25.00
Sodium Silicate	0.25
Optical Brighteners	0.25
Proteinase	0.70
Perfume	0.10
Sodium Carboxymethylcellulose	2.00
Builders and Fillers	41.70

Automatic Dish Washing Compound

FORMULA NO. 1

A	"Veegum" HS	3.0
	CMC 7M	1.0
	Water	60.0
B	"Plurafac" RA-43	2.4

"Plurafac" D-25	—
Tetrapotassium Pyrophosphate	29.2
Potassium Phosphate (tribasic)	4.4
Potassium Chloride	—

Prepare A by adding "Veegum" and CMC slowly to the water, agitating continually until smooth. Add "Plurafac" to A with stirring. Add the salts with stirring. Heat is generated by the addition of the salts. Stir until batch is cool.

No. 2

"Sodasil" P-598	25.0
Sodium Tripolyphosphate	30.0
"Plurafac" RA-43	2.0
CDB-90	1.0
Sodium Sulfate and/or Sodium Chloride	42.0

No. 3

"Unamide" L-5	4.0
Dodecyl Benzene Sulfonic Acid	16.3
Sodium Lauryl Ether Sulfate (60% active sol'n)	12.0
Sodium Hydroxide (50% sol'n)	4.3
Water	63.4

Liquid All-Purpose Cleaner Concentrate

Water	75.0
Sodium Tripolyphosphate	5.0
N Sodium Silicate	5.0
Sodium Hydroxide (50%)	5.0
"Monaterge" 85	5.0
Butyl "Cellosolve"	5.0

Use Dilution:	Light Duty	—	2-4 oz/gal
	Medium Duty	—	4-8 oz/gal
	Heavy Duty	—	16-32 oz/gal

All-Purpose Spray Cleaner

"Tergitol" 25-L-T	1.0
"Propasol" Solvent P	5.0
Tetrapotassium Pyrophosphate	2.5
Deionized Water, Perfume	q.s.

Liquid Abrasive Cleaner

A	"Veegum"	4.3
	Water	60.7
B	Calcium Carbonate (100 mesh)	30.0
C	"Plurafac" D-25	5.0
D	Perfume	q.s.
	Preservative	q.s.

Add the "Veegum" to the water slowly, agitating continually until smooth. Add the abrasive with stirring. Add the "Plurafac" and mix until smooth and uniform. Then add D.

Industrial Hand Cleaner

FORMULA NO. 1

(Gel)

"Tergitol" 25-L-T	9
"Isopar" M	35
Mineral Oil	3
Oleic Acid	6
Lanolin	1
Triethanolamine	2.1
Monoethanolamine	0.43
Propylene Glycol	3.0
Water, Dye, Perfume	q.s.

No. 2

"Maprosyl" 30	50
Bayol 90	50

Aerosol Cleaner

Formula No. 1

"Neodol" 23-6.5	1.7
Lauric Diethanolamide (LDEA)	0.5
Sodium Metasilicate-5H$_2$O	1.7
Trisodium Phosphate	1.0
Butyl "Oxitol"	3.5
Water	q.s.

Propellant used at rate of 5 parts/100 parts cleaner.

No. 2

(All-Purpose)

"Aerothene" TT Solvent	30.00
"Aerothene" MM	7.50
"Dowanol" DPM	6.00
Toluene	22.50
Diisopropanolamine	2.25
"Triton" X-100 Surfactant	1.50
Amyl Acetate	5.25
Propellant	25.00

Household Kitchen Cleaners

Formula No. 1

(With Bleach)

A	"Veegum" HS	0.50
	Water	62.85

B	Calcium Carbonate (100 mesh)	21.00
	"Super Floss"	9.00
C	Sodium Lauryl Sulfate	0.35
D	Tetrapotassium Pyrophosphate	3.00
	Potassium Phosphate Tribasic	1.50
E	Calcium Hypochlorite	1.80

No. 2

A	"Veegum" HS	1.55
	Water	68.30
B	Calcium Carbonate (100 mesh)	19.00
	"Super Floss"	8.00
C	Sodium Lauryl Sulfate	0.35
	Sodium Metasilicate (pentahydrate)	1.50
D	Calcium Hypochlorite	1.30

Add the "Veegum" HS to the water slowly, agitating continually until smooth. Add B, C, D, and E to A in order, mixing after each addition until smooth and uniform.

Borax Cleaner

Borax	98.4
Dodecylbenzene Sulfonate	0.5
Trichlorcarbonilide	0.2

All-Purpose Noncaustic Cleaner

Water	81
TKPP	10
Sodium Metasilicate (anhydrous)	5
Alkali Surfactant	2
"Triton" X-100	2

Use at ¼–½%, pH of 9.8.

Soap Detergent Bar

	FORMULA No. 1	No. 2
Dodecylbenzene Sodium Sulfonate	41-0	—
Sodium Lauryl Sulfate	15-0	—
Sodium Isethionate Coconut Acid Ester	—	47.0
Sodium Tallow-Coconut Soap (80 : 20)	30.0	—
Stearic Acid	—	30.0
Calcium Stearate	6.0	—
"Polyox" Resin WSR-35	2.0	2.0
Titanium Dioxide	1.5	1.5
Perfume	0.2	0.2
Water	4.0	4.0

Dry Cleaning Compound

FORMULA No. 1

"Maphos" 76 Na	2
Doss (70%)	1
Water	1
Perchloroethylene	96

No. 2

"Maphos" 91	2
DEA	0.3
Doss (70%)	1
Water	1
Perchloroethylene	95.7

Rug and Upholstery Shampoo

FORMULA No. 1

A	"Veegum" T	2.0
	Water	40.0

B	Sodium Lauryl Sulfate	5.0
	Water	22.4
C	Cocoyl Sarcosine	3.5
	Sodium Hydroxide (10% sol'n)	5.7
	Water	21.4

Add the "Veegum" T to the water slowly, agitating continually until smooth. Prepare B by dissolving sodium lauryl sulfate in water with agitation. Combine C and add to B. Add CB slowly to A. Mix slowly until uniform.

	No. 2	No. 3
"Hamposyl" L-30	30	40
Sodium Lauryl Sulfate (30%)	45	60
Formaldehyde (37%)	0.3	—
Water	24.7	—
Dilution for use	1 : 15	1 : 25

Spot Remover

FORMULA No. 1

Concentrate:

A	Perchloroethylene	46.7
	"Witconate" P10-59	13.0
	"Emphos" TS-230	2.4
	Triethanolamine	1.5
B	Water	31.3
	Tetrasodium Pyrophosphate	1.5
	"Emphos" CS-1361	3.6

Aerosol:

Concentrate	70.0
Propellant 12	30.0

Mix A and B separately until clear. Add B to A with agitation and mix until clear.

No. 2

Concentrate:

A	Mineral Spirits	41.60
	"Witconate" P10-59	13.35
	"Witconol" NP-60	2.85
B	Water	36.20
	Tetrasodium Pyrophosphate	1.70
	"Emphos" CS-1361	4.30

Aerosol:

Concentrate	50.00
Propellants 12/114 (35 : 65)	50.00

Mix A and B separately until clear. Add B to A with agitation and mix until clear.

This formulation demonstrates a technique whereby solvent, water, and detergent phosphate are combined in a transparent emulsion for multifunctional cleaning. This product is suggested also for treatment of difficult stains prior to laundering.

Carpet Steam Cleaner

FORMULA NO. 1

Water	80
"Petro®" 22 (50% liquid)	10
"Hampshire" NTA-150	5
Urea	2
"Tetronic" 702	3

Mix and add each item in order with mild agitation.
Use 3-5 oz/gal of water.

No. 2

Sodium Metasilicate Pentahydrate	15
Sodium Sulfate	15
Ethylene Diamine Tetraacetic Acid Tetrasodium	15
"Petro®" 22 Powder	50
"Triton" X-45	5

Blend and mix well before packaging.
Use 1 oz to 1 gal of water.

Steam Cleaner, Industrial

FORMULA NO. 1

Water	73
STPP	5
NaOH (50%)	10
42° Sodium Silicate (3.22/1)	8
Alkali Surfactant	2
"Triton" X-100	2

Use at 5 oz/gal, pH of 12.4.

NO. 2

(Degreaser)

Metasilicate (anhydrous)	36
Tripolyphosphate (light)	38
Soda Ash (light)	12
"Triton" X-100	7
"Lakeway" 301-10	7

Leather and Vinyl Cleaner

Water	95
STPP	1
NaOH (50%)	2
Alkali Surfactant	1
"Triton" X-100	1

Use at 10–15 oz/gal, pH of 12.

Automobile Vinyl Top Cleaner

A	"Veegum" HS	1.0
	CMC 7M	0.3
	Water	83.7
B	Butyl "Cellosolve"	5.0
	"Tergitol" NPX	8.0
C	Tetrapotassium Pyrophosphate	2.0

Add the "Veegum" HS-CMC dry blend to the water slowly, agitating continually until smooth. Add B and C in order to A mixing after each until smooth (avoid incorporation of air).

White Wall Tire Cleaner

Water	77.6
Sodium Tripolyphosphate	4.7
Trisodium Phosphate	4.7
"Metso" Anhydrous	3.0
"Monaterge" 85	10.0

Liquid Car Wash

"Calsoft" T-60	20
"Caloxylate" SA-9	5
"Pilot" SXS-96	2
Water	73

Glass Cleaner

FORMULA NO. 1

"Dowanol" PM	5
Propylene Glycol	5
Isopropanol	35
Water	55

NO. 2

| A | "Veegum" | 1.5 |
| | Water | 67.5 |

B	Sorbitan Monolaurate	2.0
	Polysorbate 20	2.0
	Ammonia	2.0
	Deodorized Kerosene	15.0
	Abrasive	10.0
	Preservative	q.s.

Add the "Veegum" to the water slowly, agitating continually until smooth. Add the components in B in the order listed, mixing after each addition.

Aerosol package: Concentrate 80%; Propellant 12 20%.

Window Cleaner

FORMULA NO. 1

("Fantastik ®" Type)

"Neodol" 23-6.5	1.7
Cocodiethonolamide	0.5
Trisodium Phosphate	1.0
Sodium Metasilicate $5H_2O$	1.7
Butyl "Oxitol"	3.5
Water, Color, Perfume, etc.	q.s.

NO. 2

Distilled or Deionized Water	75.0
IPA	19.4
Methyl "Oxitol"	2.0
Butyl "Oxitol"	2.5
"Neodol" 25-3A	0.1

No. 3

Distilled or Deionized Water	84.0
IPA	15.9
"Neodol" 23-6.5	0.1

Hard Surface Cleaners

	FORMULA No. 1	No. 2
	(Nonphosphate)	(Phosphate)
"Tergitol" 25-L-T	7	7
"Propasol" Solvent P	5	5
Sodium Xylene Sulfonate (actives)	4	4
EDTA, Tetrasodium Salt (actives)	2	–
Sodium Carbonate	4	–
Tetrapotassium Pyrophosphate	–	4
Trisodium Phosphate (anhydrous)	–	2
Deionized Water	to make 100	to make 100

No. 3

A	"Veegum" T	0.45
	"Kelzan"	0.15
	Water	72.40
B	"Monamid" 150-Add	0.50
	Tetrapotassium Pyrophosphate	2.00
	Water	21.00
	"Plurafac" C-17	2.50
C	Ammonium Hydroxide (28%)	1.00

Dry blend the "Veegum" T and "Kelzan" and add to the water slowly, agitating continuously until smooth. Combine B, stirring slowly to dissolve the tetrapotassium pyrophosphate (avoid incorporation of air). Add B to A with mixing. Add C with stirring.

No. 4

A	"Veegum" HS	
	"Kelzan"	
	Water	

B	"Monamid" 150-Add	0.50
	"Plurafac" C-17	2.50
	Tetrapotassium Pyrophosphate	1.25
	Potassium Phosphate (tribasic)	0,75
	Water	21.00

C	Ammonium Hydroxide (28%)	1.00

Prepare A by dry blending the "Veegum" HS and "Kelzan" and adding slowly to the water, agitating continually until smooth. Combine B, stirring slowly to dissolve the phosphates (avoid incorporation of air). Add B to A with mixing. Add C and mix until uniform.

No. 5

("Janitor In A Drum®" Type)

"Neodol" 25-9	5.0
"Neodol" 25-3	2.5
Butyl "Oxitol"	6.0
Pine Oil	0.25
Tetrapotassium Pyrophosphate	3.0
Sodium Metasilicate $5H_2O$	2.0
Sodium Xylene Sulfonate	1.0
Water	q.s.

No. 6

(All Purpose)

"Neodol" 25-3S	5
"Neodol" 25-9	5
"Neodol" 25-3	2.5
Butyl "Oxitol"	6
Pine Oil	0.25

Tetrapotassium Pyrophosphate	3
Sodium Silicate Pentahydrate	2
Sodium Xylene Sulfonate (SXS) (40% A.M.)	1
Water	q.s.

No. 7

"Unamide" C-5	5.0
NaOH (50% sol'n.)	1.0
"Hampene" 100	5.0
Sodium Metasilicate · $5H_2O$	2.0
Sodium Carbonate	1.5
Dodecyl Benzene Sulfonic Acid	3.5
Water	82.0

Wall Cleaners

FORMULA NO. 1

(Light-Duty)

"Neodol" 25-9	5
Trisodium Phosphate	2
Sodium Metasilicate $5H_2O$	2
Water	q.s.

NO. 2

(Spray)

"Neodol" 23-6.5	1.7
Coco-Acid Diethanolamine Condensate	0.5
Sodium Metasilicate $5H_2O$	1.7
Trisodium Phosphate	1.0
Butyl "Oxitol"	3.5
Water, Color, Perfume	q.s.

No. 3

(Spray)

"Neodol" 23-6.5	2.4
Tetra-Sodium Ethylenediamine Tetraacetate	2.6
Butyl "Oxitol"	3.0
Isopropyl Alcohol (99%)	1.0
Water, Color, Perfume	q.s.

No. 4

A	"Veegum" HS	1.50
	Water	62.90
B	Calcium Carbonate (100 mesh)	21.00
	"Super Floss"	9.00
C	Sodium Lauryl Sulfate	0.35
D	Tetrapotassium Pyrophosphate	3.00
	Potassium Phosphate (tribasic)	1.50
E	Calcium Hypochlorite (70% active)	0.75

Add the "Veegum" HS to the water slowly, agitating continually until smooth. Add B to A and mix until smooth. Add C, D, and E in order, mixing after each until uniform.

Toilet Bowl Cleaner, Scour

Sodium Binoxalate	1.75
Sodium Bisulfate	74.50
"ACL"	0.30
Silica	to make 100

Aluminum Soak Tank Cleaner

Water	51
NTA	2
Sodium Gluconate	2
Sodium Silicate 42° (3.22/1)	20

KOH (45%)	20
"Tomah" Alkali Surfactant	3
"Triton" X-100	2

Cost: 75 cents/gal
Use at 20 : 1 dilution.

Drain Opener

FORMULA NO. 1

Water	49
NaOH (50%)	50
Alkali Surfactant	1

NO. 2

Sodium Hydroxide	54.20
Sodium Nitrate	30.45

Egg Washing Detergent

"ACL®" 59	10–15
Sodium Tripolyphosphate	25–40
Sodium Metasilicate	0–10
Nonionic Surfactant	2–5
Soda Ash and/or Sodium Sulfate	to make 100

Dishwashing Detergent

FORMULA NO. 1

Sodium Tripolyphosphate	30–45
Water	4–10
Nonionic Surfactant	1–3

Sodium Carbonate	5-15
Sodium Metasilicate or Hydrous Silicate	10-15
Sodium Sulfate (if required as filler)	to make 100
"ACL®" 56	1-1.5

Mixing is generally accomplished in a ribbon, paddle or tumble type blender. The order of mixing of the components should be such that the nonionic surfactant is adsorbed onto the alkaline chemicals prior to addition of the chlorinated s-triazine triones, which should be added last.

No. 2

Sodium Tripolyphosphate	30-45
Nonionic Surfactant	1-3
Sodium Carbonate	10-25
Sodium Sulfate	to make 100
"ACL®" 56	1-1.5
Silicate Solution (40-50% solids)	20-30

Scouring Powder

"ACL®" 59	0.5-0.7
Trisodium Phosphate or Soda Ash	3-5
Anionic Surfactant	2-5
Silica Flour	to make 100

Furniture Cleaner

Concentrate:

Mineral Oil	20.0
"Witconol" 14	0.5
Water	79.5

Aerosol:

| Concentrate | 60.0 |
| Propellant 12 | 40.0 |

Valve:

    Vapor-tap valve with capillary dip-tube, mechanical breakup actuator.

    Dissolve "Witconol" 14 surfactant in mineral oil. Slowly pour in water while stirring to effect emulsification.

Floor Cleaner

FORMULA No. 1

"Macol" NP 10	7
STPP	8
TSP	2
Sodium Metasilicate · 5H$_2$O	2
"Maphos" 80	2-3
Water	81

No. 2

"Dowanol" PM	14
Triethanolamine	12
Oleic Acid	16
Water	58

No. 3

"Mazamide" 65	8
Sodium Tripolyphosphate	2
Tetrapotassium Pyrophosphate	2
Water	88
Dye and Perfume to suit	—

Garage Floor Cleaner

"Neodol" 23-6.5	2.5
Tallow Soap	2
Butyl "Oxitol"	6.5

Trisodium Phosphate	3
Sodium Metasilicate · 5H$_2$O	3
Water	q.s.

The following mixing sequence is recommended: Stir "Neodol" 23-6.5 and Butyl "Oxitol" to homogeneous mixture, add tallow soap then water. When the soap is dissolved add trisodium phosphate and sodium metasilicate.

The garage floor cleaner is applied directly to oil spots. After allowing 5 min for penetration, flush with garden hose.

Wax Stripper

Water	89
Na$_3$ NTA	1
Sodium Metasilicate · 5H$_2$O	5
Alkali Surfactant	2
Monoethanolamine	3

Use 8 oz/gal, pH of 12.0.

Paint Stripper

Water	40
Sodium Gluconate	1
45% KOH	55
Alkali Surfactant	4

Water Softener

| Sodium Hexametaphosphate | 85 |
| Water | 15 |

Sweeping Compound

Benzaldehyde	1
Paradichlorbenzene	1
Mineral Oil	26

Wood Flour	25
Sand, Fine	41
Color	to suit
Water	6

Copper Cleaner

FORMULA NO. 1

A	"Veegum" K	2.07
	"Kelzan"	0.23
	Water	78.55
B	"Snow Floss"	13.60
C	Buffer Solution†	q.s.
D	"Triton" X-102	4.65
	Lauryl Thioglycolate	0.90
	Perfume and Preservative	q.s.

†Buffer solution: 1.46 parts—1M H_3PO_4
            1 part—125 grams/liter Na_3PO_4

Dry blend "Veegum" and "Kelzan" and add to the water slowly, agitating continually until smooth. Add B to A gradually. Mix until smooth, then buffer this mixture with buffer solution C to a pH of 2.5. Mix components in D until a clear solution is formed. Special care should be taken to avoid incorporation of air. Add D to other components very slowly and mix until uniform.

Directions for use: Apply copper cleaner with damp cloth. Rinse and dry. Polish with a clean dry cloth.

NO. 2

A	"Veegum" K	2.07
	"Kelzan"	0.23
	Water	78.55
B	"Snow Floss"	13.60
C	Buffer Solution†	q.s.

D	"Triton" X-102	4.65
	Benzotriazole	0.90
E	Perfume and Preservative	q.s.
	Color	q.s.

†Buffer solution: 1.46 parts—1M H_3PO_4
1 part—125 grams/liter Na_3PO_4

Dry blend "Veegum" K and "Kelzan" and add to the water slowly, agitating continually until smooth. Add B to A gradually. Mix until smooth, then buffer this mixture to a pH of 2.5. Mix components in D until a clear solution is formed. Special care should be taken to avoid incorporation of air. Add D to other components very slowly and mix until uniform. Then add E.

Car Washes

FORMULA NO. 1

(High Viscosity Liquid)

"Neodol" 23-6.5	15
"Neodol" 25-3A	5
"Natrosol" 250 HR	1
Water	q.s.

NO. 2

(Liquid)

"Neodol" 23-6.5	3
Sodium Tripolyphosphate	27
Water	q.s.

NO. 3

(Solid Concentrate)

"Neodol" 23-6.5	20
Sodium Carbonate	40
Sodium Pyrophosphate	40

Institutional Oven Cleaner

Water	59
Sodium Gluconate	as desired
NaOH Flake	20
Alkali Surfactant	10
Propylene Glycol	6
Butyl "Cellosolve"	4
"Triton" X-100	1

Use as is, pH of 14.

Aerosol Protective Oven Film Cleaner

A	"Veegum" T	2.9
	Water	86.4
B	"Pluronic" F-127	4.3
C	"Dow Corning" 200 Fluid (60,000 cs)	6.4

Add the "Veegum" T to the water slowly, agitating continually until smooth. Add B to A with agitation. Add C, mixing until uniform.

Concentrate: 70%; propellant 12: 30%.

Directions for use: Spray evenly on clean unheated oven walls from a distance of about 10 in. After cooking allow oven to cool, then wipe off film and grease splatter with damp sponge.

Waterless Hand Cleaner

A	Mineral Oil (light)	20.00
	Deodorized Kerosene	20.00
	Stearic Acid (triple pressed)	5.00
	"Lexemul" 515	3.00
	"Lexgard" P	0.10
B	Water	45.20
	Propylene Glycol	5.00
	Triethanolamine	1.50
	"Lexgard" M	0.20

Heat A to 60 C. Heat B to 65 C. Add A to B and blend until cool to 35 C. Fill.

Industrial Sanitizer-Cleaner

"BTC®" 2125M (50%)	3.2
Sodium Metasilicate (anhydrous)	2.4
"Perma Kleer" 100	2.5
"Neutronyx" 656	5.0
Tetrapotassium Pyrophosphate (anhydrous)	5.0
Water	81.9

Institutional Disinfectant Cleaner

"BTC®" 2125M (50%)	9.0
Sodium Carbonate	3.0
"Neutronyx" 656	4.5
Sodium Tripolyphosphate (anhydrous)	2.0
Water	81.5

Hospital Disinfectant Cleaner

"BTC®" 2125M (50%)	18.0
Sodium Carbonate	4.0
"Perma Kleer" 100	5.0
"Neutronyx" 656	9.0
Water	64.0

Dairy Cleaner-Sanitizer

"BTC®" 2125M (50%)	20.0
"Neutronyx" 656	10.0
Phosphoric Acid (85%)	35.3
Water	34.7

Detergent Disinfectant, Aerosol

"BTC®" 2125M (50%)	0.40
"Perma Kleer" 100	4.00
"Neutronyx" 656	0.50
Ethylene Glycol Monobutyl Ether	2.00
Diethylene Glycol Monoethyl Ether	2.00
Sodium Metasilicate · $5H_2O$	0.25
Pine Needles Perfume SYN 802277	0.20
Water	85.65
Isobutane Propellant	5.00

Disinfectant Cleaner

FORMULA NO. 1

"Veegum"	2.0
Water	89.0
"Super Floss"	1.6
Sodium Hypochlorite (5% active Cl)	5.0

NO. 2

"ACL®" 85	4
Sodium Tripolyphosphate (anhydrous)	40
Sodium Sulfate (anhydrous)	56

NO. 3

"Bardac"-20 (50% active sol'n)	3.80
"Hamp-ene" 100 (40% active sol'n)	2.50
Trisodium Phosphate	2.00

Inert Ingredients:	
Tetrapotassium Pyrophosphate	4.00
Sodium Sesquicarbonate	2.00
"Tergitol" TP-9	4.00
Water	81.70

Add the "Tergitol" TP-9 to the water, and stir until the surfactant has been completely dissolved. Follow with the addition of "Hamp-ene" 100 and "Bardac"-20. Stir until solution is clear. During continued stirring add tetrapotassium pyrophosphate, trisodium phosphate, and sodium sesquicarbonate. Continue stirring until the salts are completely dissolved.

No. 4

(Phosphate-Free)

"BTC®" 2125M (50%)	6.4
Sodium Carbonate	3.0
"Perma Kleer" 100	2.5
"Neutronyx" 656	4.5
Water	83.6

	No. 5	No. 6
"Bardac" 22	5	10
"Neodol" 25-9	5	7.5
Tetrapotassium Pyrophosphate	4	8
Trisodium Phosphate	2	—
Butyl "Oxitol"	5	—
Water	79	74.5

Deodorizer Sanitizer

Formula No. 1

"BTC®" 2125M (50%)	20
Water	80

No. 2

"ACL®" 59	20–25
Sodium Tripolyphosphate	0–15
Sodium Sulfate	to make 100

Aerosol Disinfectant

"BTC®" 2125M (50%)	0.40
Triethylene Glycol	6.00
Isopropanol	53.00
Water	10.00
Sodium Nitrite	0.10
Perfume	0.50
Propellant	30.00

Deodorizer, Cesspool

Parachlorometaxylenol	5
Potassium Ricinoleate	6
Isopropanol	10
Terpineol	10
Water	to 100
Ph	9.0–9.5

Deodorizing Toilet Drip Fluid

Kerosene	45
Mineral (Seal) Oil	45
Eucalyptus Oil	3
Nitrobenzol	5
Benzaldehyde	2

Latrine Cleaner

Hydrochloric Acid	52.0
o-Dichlorobenzene	3.0
"Igepal" CO-630	2.0
"Igepal" CO-530	0.5
Water	42.5

Chapter VI

DRUG PRODUCTS

Analgesic

FORMULA NO. 1

(Internal)

A	"Veegum" K	1.0
	Water	90.6
B	Zinc Sulfocarbolate	0.4
	Phenyl Salicylate	1.0
	Bismuth Subsalicylate	3.0
C	Ethanol	4.0
D	Preservative	q.s.

Add the "Veegum" to the water slowly, agitating continually until smooth. Add B to A slowly with mixing. Add C and stir until smooth. Add D.

No. 2

(Cream)

A	"Veegum" PRO	1.75
	"Klucel" J	1.75
	Water	40.75
	Ethanol	36.75

B	Eucalyptus Oil	0.20
	Menthol	0.40
	Peppermint Oil	0.40
	Triethanolamine	3.00
	Methyl Salicylate	5.00
	Cocoyl Sarcosine	10.00
	Preservative	q.s.

Dry blend the "Veegum" PRO and "Klucel" J and add to the water and ethanol slowly, agitating continually until smooth. Combine B, stirring to dissolve crystals, and add to A with slow mixing. Mix until uniform.

No. 3

(Heat-Rub)

Menthol USP	10.0
Methyl Salicylate	15.0
"Crodafos" N.3 Neutral	2.5
"Klucel" HP (4)	2.5
Water	30.0
Ethanol SDA-40 (95%)	40.0

Use a slow speed anchor mixer that can develop a high torque. First heat the distilled water to 50–60 C, and add the "Klucel" slowly, with stirring. Continue stirring and let the mixture cool to room temperature (it will form a heavy gel). Next, add the ethanol and mix until uniform. Combine the remaining ingredients and add them slowly. Stir the formulation until fairly uniform, then fill off. The gel will be grainy at first, but becomes smooth on standing.

No. 4

(Water Rinsable)

A	"Ekaline" G Flakes	20.00
	Methyl Salicylate USP	30.00

	Menthol	5.00
	Isopropyl Myristate	2.50
B	Water	42.00
	Lactic Acid (88%)	0.09
	"Chemical Base" 6532	0.30
	Preservatives and misc.	q.s.

Separately heat A and B to 65 C with stirring. When both are homogeneous add B to A with stirring. Cool with stirring to 38 C.

No. 5

(Internal)

Aspirin USP	33.44
Salicylamide	16.72
Acetaminophen USP	16.72
Caffeine USP (granular)	5.60
"Avicel" PH-101	25.00
Stearic Acid (powder)	2.00
"Cab-O-Sil"	0.52

Blend all the ingredients, except the stearic acid, for 25 min. Screen in the stearic acid and blend for an additional 5 min. Compress.

Burn Cream

FORMULA No. 1

A	"Lexemul" 530	10.00
	"Lexol" IPM	12.00
	Methyl Salicylate	5.00
	Benzocaine	2.20
	Stearic Acid (triple pressed)	2.00
	Menthol	0.70
	Camphor	0.40
B	Water	61.40
	Propylene Glycol	5.00

Triethanolamine	1.00
"Versene" 100	0.10
"Lexgard" M	0.20

No. 2

(Antiburn Jelly)

"Manucol" KMR	2.7
Glycerin	3.0
Methyl p-Hydroxybenzoate	0.1
Proflavine	0.1
Water	to 100

Disperse the proflavine in the water. Add the "Manucol" KMR to the glycerin and preservative. Add the "Manucol" KMR dispersion to the water with high shear mixing. Continue stirring until dissolved.

Suppository Base

"Crodamol" SS	16.7
"Super Corona" Lanolin	16.7
Paraffin Wax 125	21.0
Mineral Oil	45.6

Heat all ingredients together at 65 C.

Acne Cream

FORMULA No. 1

A	"Lexemul" 515	6.00
	Stearic Acid (triple pressed)	2.00
	"Lexol" IPM	3.00
B	Water	64.80
	Propylene Glycol	3.00
	Resorcinol	3.00
	Triethanolamine	1.00
	"Lexgard" M	0.20

C	Sulfur (precipitated)	2.00
	Bentonite	5.00
	Ethanol (95%)	10.00

Disperse the sulfur into B. Disperse the bentonite into B while it is heating. Heat A and B to 80 C. Add A to B with agitation. Cool with agitation until the emulsion begins to get heavy and smooth (about 50 C). Very slowly and with very good agitation add the alcohol. Continue cooling to about 30 C.

No. 2

"Pluronic" F-127 Polyol	20
Sulfur Dispersion	3
Water	77
Perfume	q.s.
Preservative	q.s.

Place cold water (5-10 C) in container and add "Pluronic" F-127 polyol slowly with good agitation. Mix until all F-127 is dissolved, maintaining temperature at 10 C. Add sulfur fluid, perfume and preservative; mix until homogeneous and transfer to containers while still cool. The mixture sets up into a crystal clear, yellow ringing gel at room temperature.

No. 3

A	"Veegum" HV	2.00
	Water	38.35
	"Methocel" E4M	1.50
B	Alcohol SDA-40	30.00
	Propylene Glycol	6.00
	"Brij" 30	5.00
C	Benzoyl Peroxide (70%)	7.15
	Water	10.00

Add the "Veegum" HV to the water slowly, agitating continually until smooth. Add the "Methocel" and mix briefly to obtain a uniform slurry. Add B to A and mix until smooth ("Methocel" dissolved). Micro-pulverize C and add to A and B, mixing until smooth and uniform. This formula is a gel of solid cream consistency and may best be dispensed from a standard wide mouth jar.

Psoriasis Cream

FORMULA NO. 1

"Super Hartolan" Lanolin	4.0
Petrolatum	5.0
Paraffin Wax 140	12.0
Mineral Oil	30.0
Coal Tar	2.0
Allantoin	0.25–0.5
Water	q.s. to 100

Heat both phases to 70 C. Add the water slowly to the oil phase with high speed stirring, allowing time for each addition to incorporate. When all in, cool with agitation to just above melting point and fill off. A homogenizer may be used to improve stability.

NO. 2

Stearic Acid (triple pressed)	16.0
"Novol"	6.0
Lanolin	2.0
Coal Tar	2.0
Triethanolamine	0.6
Allantoin	0.25–0.5
Water	q.s. to 100

Heat both phases to 75–80 C, add water to oils with high speed stir-ring. Cool to 40 C with stirring and fill off.

Cortisone Cream

"Pluronic" F-127	20.0
Hydrocortisone	0.5
Neomycin Sulfate	0.5
Distilled Water	79.0

In a container equipped with a stirrer, melt the "Pluronic" F-127 polyol and hydrocortisone together (55-60 C). Add cold water (5-10 C) and continue mixing slowly, keeping the mixture cool (< 10 C). Add neomycin sulfate when all lumps are gone. Mix for a few minutes, and transfer to suitable containers. Product sets up into an opaque cream at room temperature.

Ophthalmic Gel

"Pluronic" F-127	20.00
Boric Acid	1.50
Zinc Sulfate	0.25
Benzethonium Chloride	0.02
Distilled WAter	78.23

In a suitable vessel equipped with a stirrer, weigh out boric acid, zinc sulfate, the benzethonium chloride and cold (5-10 C) water. Mix gently and add the "Pluronic" F-127 polyol slowly while continuing to mix at 5-10 C. When all ingredients have dissolved, transfer to containers and allow to warm to room temperature. The product forms a clear gel.

Antiseptic Gel

FORMULA NO. 1

"Tetronic" 1508 Polyol	26
Propylene Glycol	2
"Roccal"	1
Water	71

No. 2

"Pluronic" F-127	20
Propylene Glycol	2
"Roccal"	1
Water	77

Dissolve surfactant in cold water (5-10 C). When homogeneous, add propylene glycol. (Gel strength may be altered by varying the amount of surfactant.) Product sets up into a clear, ringing gel when warmed to room temperature.

Heat Rub

"Crodafos" SG	2.5
"Polawax"	4.0
"Procetyl" 10	4.5
Menthol	1.0
Camphor	1.0
Methyl Salicylate	7.5
Glycerin	3.0
"Crodyne" BY19	1.0
Diethanolamine	0.3
Preservative	0.1
Water	75.1

Heat oils to 60 C. Dissolve "Crodyne" in the water. Heat water to 65 C. Add water to oils with high stirring. When uniform cool to 35 C and fill off. Lotion will thicken on standing.

Medicated Rubs

	FORMULA No. 1	No. 2
	(Cream)	*(Lotion)*
Stearic Acid (triple pressed)	16.0	10.0
Lanolin USP	4.0	—
"Lanexol" AWS	—	2.0
"Crodafos" SG	—	2.0

Medication Mix*	0.3	0.3
Propyl Paraben	0.1	0.1
Methyl Paraben	0.1	0.1
"Crotein" SPC	0.5	0.5
Triethanolamine	0.5	2.0
Limewater	30.0	—
Distilled Water	47.2	83.0

Heat the oils to 75 C and stir until the solid ingredients are dissolved. Heat the water phase and aqueous soluble ingredients to 75 C. Cool with stirring to 45 C when the medication mix is added. The lotion should be filled off while liquid.

*Medication Mix: 15% menthol, 30% camphor, 10% clove oil, 15% eucalyptol, and 30% phenol.

Surgical Scrub

Hexachlorophene	3.0
Propylene Glycol	5.0
"Cremba"	2.5
"Triton" X-200	70.0
Water	19.5

Dissolve the hexachlorophene in the propylene glycol with heating to 55 C. Add the "Cremba." Heat the "Triton" and water to 60 C. Add this in a thin stream to the other phase with high speed agitation. When all in, continue agitation with cooling to 30 C and fill off.

Antacid

FORMULA NO. 1

A	"Veegum"	0.8
	Sodium Carboxymethylcellulose (low visc.)	0.6
	Water	32.8
B	Sorbo	16.8
C	Magnesium Hydroxide	3.2

Aluminum Hydroxide	3.2
Saccharin	0.2
D Water	42.4
E Preservative	q.s.

Dry blend the "Veegum" and CMC and add to the water slowly, agitating continually until smooth. Add B to A and mix until uniform. Add C to A and B slowly with mixing. Add D and homogenize. Then add E.

NO. 2

Reheis F-MA 11	325.0
Mannitol USP (granular)	675.0
"Avicel®"	75.0
Starch	30.0
Calcium Stearate	22.0
Flavor	q.s.

Blend all ingredients and compress using a 5/8 in. flat face bevel-edge punch to a hardness of 8–11 kg (Strong-Cobb-Arner tester).

NO. 3

	Per Tablet
Aluminum Hydroxide	325.0 mg
Mannitol USP (powdered)	812.0 mg
Sodium Saccharin	0.4 mg
Sorbitol (as 10% W/V sol'n)	32.5 mg
Magnesium Stearate	35.0 mg
"Felcofix®" Mint Concentrate	4.0 mg

Blend the antacid, mannitol, and sodium saccharin. Granulate with the sorbitol solution. Dry at 120 F and screen through 12-mesh screen. Add flavor and magnesium stearate. Blend and compress on 5/8 in. flat face bevel-edge punch to a hardness of 8–11 kg (Strong-Cobb-Arner tester).

No. 4

A	"Veegum" HS	2.00
	Water	47.06
B	Aluminum Hydroxide Gel AHLT-LW	44.50
	Sodium Saccharin	0.03
	Oil of Peppermint	0.01
	Sorbitol	6.40
	Preservative	q.s.

Add the "Veegum" HS to the water slowly, agitating continually until smooth. Combine the B ingredients, mixing until smooth. Add B to A with stirring, and mix until smooth.

Triple Sulfa Suspension

A	"Veegum"	1.00
	Sugar Syrup	90.60
B	Sodium Citrate	0.78
C	Sulfadiazine	2.54
	Sulfamerazine	2.54
	Sulfamethazine	2.54
D	Citric Acid (5% sol'n.)	q.s.
	Preservative	q.s.
	Flavor and Color	q.s.

Prepare sugar syrup, add the "Veegum" slowly, agitating continually until smooth. Add B to A and mix. Dry blend C and add to A and B. Mix until smooth. Homogenize. Buffer to pH 5.6 with D. Remix the next day.

Kaolin-Pectin Suspension

A	"Veegum"	0.88
	Cellulose Gum	0.22
	Water	79.12

B	Kaolin	17.50
C	Pectin	0.44
	Saccharin	0.09
	Glycerin	1.75
	Flavor	q.s.
	Preservative	q.s.

Dry blend "Veegum" and cellulose gum and add to the water slowly, agitating continually until smooth. Add B and mix. Mix C and add to other components.

Vitamin E Acetate Clear Dispersion

"Tween 80"	6
Vitamin E Acetate (aqueous)	2
Glycerin	12

Vitamin E Tablet

Dry Vitamin E Acetate	80
"Syloid" 74	1
"Avicel" PH-102	

Blend and compress.

Vitamin B$_1$ Tablet

Thiamine Hydrochloride USP	30.0
"Avicel" PH-102	25.0
Lactose (anhydrous)	42.5
Magnesium Stearate	2.0
"Cab-O-Sil"	0.5

Blend all ingredients, except the magnesium stearate, for 25 min. Screen in the magnesium stearate and blend for an additional 5 min. Compress.

Multivitamin Tablet

(Maintenance)

1.	Vitamin A Acetate (dry form 500 UA and 500 U D_2 per mg)	10
2.	Vitamin A Acetate USP	10
3.	Vitamin D_2	10
4.	Thiamine Mononitrate USP	10
5.	Riboflavin USP	10
6.	Pyridoxine HCl USP	5
7.	"Stabicote"	10
8.	D-Calcium Pantothenate USP	50
9.	Ascorbic Acid USP fine powder for tableting	10
10.	Ascorbic Acid	10
11.	Niacinamide	10
12.	Dicalcium Phosphate (unmilled)	
13.	"Avicel" PH-101	
14.	Talc USP	
15.	Stearic Acid	
16.	Magnesium Stearate	

Roughly blend ingredients 4, 5, 6, 8, 9, 13, 14, 15, and 16 and mill using the Fitzmill with No. 2A screen, impact forward and at medium speed. Add ingredients 1, 2, 3, 7, 10, 11, and 12 through a No. 30 screen and blend thoroughly. Compress at a tab. weight of 200.0 mg using a Rotary tablet machine with punch and die assembly.

Multivitamin Tablet

FORMULA NO. 1

(Slugging)

	Per Tablet	
Vitamin A (coated)	5000	USP units
Vitamin D (coated)	400	USP units
Vitamin C (ascorbic acid) (coated)	60.0 mg	
Vitamin B_1 (thiamine mononitrate)	1.0 mg	
Vitamin B_2 (riboflavin)	1.5 mg	

Vitamin B_6 (pyridoxine hydrochloride)	1.0 mg
Vitamin B_{12} (cyanocobalamin)	2.0 μg
Calcium Pantothenate	3.0 mg
Niacinamide	10.0 mg
Sodium Saccharin	1.1 mg
Flavor	q.s.
Mannitol USP (powdered)	236.2 mg
Starch	16.6 mg
Magnesium Stearate	6.6 mg
Talc	10.0 mg

Blend all ingredients, slug and pass through a 20-mesh screen. Compress on 3/8 in. standard concave punch to a hardness of 7–8 kg (Strong-Cobb-Arner tester).

No. 2

(Wet Granulation)

	Per Tablet	
Vitamin A (coated)	5000	USP units
Vitamin D (coated)	400	USP units
Vitamin C (ascorbic acid) (coated)	60.0 mg	
Vitamin B_1 (thiamine mononitrate)	1.0 mg	
Vitamin B_2 (riboflavin)	1.5 mg	
Vitamin B_6 (pyridoxine hydrochloride)	1.0 mg	
Vitamin B_{12} (cyanocobalamin)	2.0 μg	
Calcium Pantothenate	3.0 mg	
Niacinamide	10.0 mg	
Sodium Saccharin	0.3 mg	
Flavor	q.s.	
Mannitol USP (powdered)	234.8 mg	
Acacia Powder	6.5 mg	
Magnesium Stearate	6.5 mg	
Talc	7.0 mg	

Blend the mannitol, saccharin and acacia with 10% of the riboflavin and all other vitamins except A, D and C. Granulate this blend with water. Dry at 120 F, pass through a 16-mesh screen and add flavor. Mix ascorbic

acid with magnesium stearate, and mix vitamins A and D with remainder
of riboflavin. Add these and the talc to the previous mixture and mix well.
Compress on 3/8 in. concave punch to a hardness of 7-8 kg (Strong-Cobb-
Arner tester).

Cough-Cold Tablets

(Wet Granulation)

	Per Tablet
Dextromethorphan Hydrobromide (15 mg per tablet)	
as 10% Adsorbate	150.0 mg
Mannitol USP	350.0 mg
"Avicel" PH	25.0 mg
Citric Acid	15.0 mg
Sodium Saccharin	1.1 mg
Binder Solution* (add in proportion of 235 ml for	
1000 tablets)	
Calcium Stearate	10.8 mg
Orange Juice DF FOL MM 3X 273	4.0 mg

*Sorbitol (powdered)	200 g
Glycerin (99.5%) USP	10 g
Water q.s. ad	1000 ml

Screen dextromethorphan hydrobromide adsorbate, mannitol, "Avicel" PH, citric acid, and sodium saccharin through 40-mesh screen. Mix
well. Granulate and dry overnight at 140-150 F. Screen through 12-mesh
screen. Add flavor and calcium stearate. Mix well. Compress on 7/16 in.
standard concave punch to a hardness of 5-7 kg (Strong-Cobb-Arner tester).

Vitamin C Tablet

(Chewable)

Ascorbic Acid USP	12.26
Sodium Ascorbate USP	36.26
"Avicel" PH-101	17.12
Sodium Saccharin (powder)	0.56
Sucrose	29.30

Stearic Acid (fine powder)	2.50
Imitation Orange Juice Flavor	1.00
FD & C Yellow No. 6 Dye	0.50
"Cab-O-Sil"	0.50

Blend all ingredients, except the stearic acid, for 25 min. Screen in the stearic acid and blend for an additional 5 min. Compress.

Aspirin Tablet

(5 Grain)

Aspirin USP	80.0
"Avicel" PH-101	12.0
Corn Starch USP	7.0
Stearic Acid (powder)	1.0

Blend all the ingredients, except the stearic acid, for 25 min. Screen in the stearic acid and blend for an additional 5 min. Compress.

Children's Aspirin Tablets

	Per Tablet
Aspirin (20-mesh)	81.0 mg
Mannitol USP (powdered)	1000 g
Sodium Saccharin	7 g
10% Acacia Sol'n.	230 ml
Starch	10.6 mg
Talc	8.0 mg
Stearic Acid	0.3 mg
Flavor	q.s.

Granulate the mannitol and saccharin with the acacia solution, dry at 110 F and pass through a 20-mesh screen.

Mix the aspirin with the dried mannitol granulation. Take extreme care to avoid moisture pick-up. Add the flavor to the starch, mixing thoroughly. Add the talc and stearic acid, screened to 100-mesh, to the flavor-starch mixture. Blend this combination of flavoring, starch, talc and stearic acid with the mannitol-aspirin mixture.

Compress using 11/32 in. standard concave punch to a hardness of 4-6 kg (Strong-Cobb-Arner tester).

Phenobarbital Tablet

1.	Phenobarbital	23
2.	"Avicel" PH-101	23
3.	Lactose	52
4.	"Quso" F-22	1
5.	Stearic Acid	1

Blend 1 and 2, add 3, blend, add 4 and 5, blend; compress.

Acetaminophen Chewable Tablet

	Per Tablet
Mannitol USP (powdered)	720.0 mg
Sodium Saccharin	6.0 mg
Acetaminophen NF	120.0 mg
Binder solution*	21.6 mg
Peppermint Oil	0.5 mg
"Syloid®" 244	0.5 mg
Banana, "Permaseal®" F-4932	2.0 mg
Anise, "Permaseal®" F-2837	2.0 mg
Sodium Chloride (powdered)	6.0 mg
Magnesium Stearate	27.4 mg

*Prepare binder solution consisting of:

Acacia (powdered)	15 g
Gelatin (granular)	45 g
Water q.s. ad	500 ml

Screen mannitol and sodium saccharin through 40-mesh screen. Blend thoroughly with the acetaminophen. Granulate and dry overnight at 140-150 F. Screen through 12-mesh screen. Adsorb the peppermint oil onto the "Syloid" 244 and mix with the flavors and sodium chloride. Blend this flavor mixture, the dried granulation and magnesium stearate. Compress on ½ in. flat face bevel edge punch to a hardness of 12–15 kg (Strong-Cobb-Arner tester).

PAS Tablet

Sodium Aminosalicylate USP	60.0
"Avicel" PH-102	19.5
"STA-Rx" 1500	19.0
Stearic Acid (powder)	1.0
"Cab-O-Sil"	0.5

Blend the ingredients, except the stearic acid, for 25 min. Screen in the stearic acid and blend an additional 5 min. Compress.

Chlorpromazine Tablet

Chlorpromazine Hydrochloride USP	28.0
"Avicel" PH-102	35.0
Dicalcium Phosphate (unmilled)	35.0
"Cab-O-Sil"	0.5
Magnesium Stearate	1.5

Blend all the ingredients, except the magnesium stearate for 25 min. Screen in the magnesium stearate and blend for an additional 5 min. Compress.

Penicillin V-Potassium Tablet

Penicillin V Potassium USP	50.00
"Avicel" PH-102	24.25
Dicalcium Phosphate (anhydrous)	22.00
Magnesium Stearate	3.75

Blend the penicillin V potassium, "Avicel" PH-102, and dicalcium phosphate for 25 min. Screen in the magnesium stearate and blend for an additional 5 min. Compress.

Isosorbide Dinitrate Tablet

Isosorbide Dinitrate (25% in lactose)	20.00
"Avicel" PH-102	19.80
Lactose (spray-dried)	59.45
Magnesium Stearate	0.75

Weigh all ingredients. Blend for 30 min in a P-K blender. Compress.

Allergy Relief

Phenyl Propanolamine HCl	75
Chlorpheniramine Maleate	8

Acne Cream

FORMULA NO. 1

A	"Veegum"	1.75
	Cellulose Gum	0.40
	Water	34.70
B	Glycerin	5.00
	Allantoin	0.25
	Resorcinol	3.00
	"Triton" X-100	0.20
	Water	29.70
C	"Nytal" 300	16.00
	Titanium Dioxide	2.90
	Iron Oxides	1.10
	Sulfur	5.00
	Preservative	q.s.

Dry blend the "Veegum" and CMC and add to the water slowly, agitating continually until smooth. Add B to A. Pulverize C and add to A and B. This formula is a soft cream.

No. 2

(Scrub Cream)

A	"Veegum"	2.0
	Water	58.2
B	Propylene Glycol	10.0
	"Amerchol" L-101	15.0
	"Aldo" MSA	3.0
	"Triton" X-202	1.4
C	"A-C" 9A Polyethylene	10.0
D	Eucalyptus Oil	0.4
	Preservative	q.s.

Prepare A by adding the "Veegum" slowly to the water, agitating continually until smooth. Heat to 75 C. Heat B ingredients to 70 C. Add B to A with mixing until smooth. Add C and D with mixing until cool.

No. 3

(Biostatic Lotion)

A	"Veegum" K	1.50
	Water	34.00
B	Glycerin	5.00
	Salicylic Acid	3.00
	"Triton" X-202	0.70
	Water	30.65
C	"Pyrax" B	16.00
	Titanium Dioxide	2.90
	Iron Oxides	0.25
	Sulfur	5.00
D	"Vancide" 89RE	1.00

Add the "Veegum" K to the water slowly, agitating continually until smooth. Add B to A and mix until uniform. Mix C together. Wet the powders with a portion of A and B and then work in the remainder to form a smooth lotion. Add D and mix until uniformly dispersed.

Antivy Lotion

A	"Veegum"	1.50
	Sodium Carboxymethylcellulose (med. visc.)	0.30
	Water	79.65
B	Zirconium Oxide	4.00
	Propylene Glycol	5.00
C	Isopropyl Alcohol	8.00
	Benzocaine	1.50
	Menthol	0.05
	Preservative	q.s.

Dry blend the "Veegum" and CMC and add to the water slowly, agitating continually until smooth. Add B to A. Add C to A and B with rapid mixing.

Ear Wax Dissolver

Glycerin	10
Urea Peroxide	1

Warm and mix.

Foot Powder, Antiseptic

FORMULA NO. 1

Boric Acid	18.0
Salicylic Acid	2.0
Zinc Sulfate	6.0
Zinc or Aluminum Phenolsulfonate	4.0
Talc	70.0
Perfume	to suit

Place all ingredients together in a powder mixer and grind until all materials are uniformly mixed.

No. 2

1.	Talc		82.65
2.	"Oat-Pro®"		3.00
3.	"Microdry"		10.00
4.	"Syloid" 72		2.00
5.	"Ottasept" Extra		0.15
6.	Zinc Oxide		2.00
7.	Perfume		q.s.

Add 2-7 to 1 and blend until completely uniform.

No. 3

1.	Talc	q.s.	100.00
2.	"Oat-Pro®"		10.00
3.	"Microdry"		5.00
4.	Zinc Oxide		2.00
5.	"Syloid" 72		2.00
6.	Perfume		q.s.

Add 2-6 to 1 and blend until uniform.

Athlete's Foot Remedy

Poloxamer 407	16
Undecylenic Acid	5
Isopropyl Alcohol	20
Water	59

Baby Powder

1.	Talc	q.s.	100.00
2.	Magnesium Stearate		2.00
3.	"Syloid" 72		1.00

4. "Ottasept" Extra	0.10
5. "Oat-Pro®"	10.00
6. Perfume	q.s.

Add ingredients 2–6 to the talc; blend until uniform.

Corn Remover

Salicylic Acid	12
Benzoic Acid	6
"Poloxamer" 407	47
Water	35
Perfume and Color	q.s.

Callus Softener

Salicylic Acid	3.0
Phenol	0.5
Water	10.0
Alcohol	86.5

Coal Tar Ointment

Crude Coal Tar	3
Zinc Oxide	3
Starch	27
White Petrolatum	q.s. ad 60

Calamine Emulsion

Prepared Calamine	9.6
Zinc Oxide	9.6
Cottonseed Oil	60.0
Lime Water	q.s. ad 120.0

Slimming Food Drink

"Manucol" LMR	24.0
Calcium Citrate	4.0
Sodium Citrate	12.0
Lactose or Glucose	20
Color, Flavor, Saccharin	as required

Essential vitamins, minerals, and powdered milk are blended in.

Barium Sulfate Suspension

"Manucol" LMR	0.23
Barium Sulfate	50.0
Water	85.0

Mouth Wash

FORMULA NO. 1

Alcohol	24.5
Menthol	0.1
"Tween" 60	1.0
"Tween" 20	0.4
Phenyl Salicylate	0.2
Zinc or Aluminum Phenolsulfonate	0.4
Water	73.0

Dissolve the phenolsulfonate salt in the water. Dissolve the menthol and phenyl salicylate in alcohol. Add "Tween" to alcoholic solution. Add alcoholic solution to the aqueous solution slowly with continuous stirring. Age the formulation for a sufficient length of time to permit all the insoluble material to be precipitated. Decant the supernant liquid and cool to 40 C. Filter while cold.

No. 2

Alcohol			18.0
Zinc or Aluminum Phenolsulfonate			3.0
Glycerin			8.0
Menthol Anethole Cassia Sassafras Ginger Clove	Equal parts	prepare mixture of each, then use	1.0
"Tween" 20			2.0
Water			67.8

Dissolve the phenolsulfonate salt, menthol and oils in a mixture of glycerin, "Tween" 20 and alcohol. Add water to alcoholic solution slowly and with continuous stirring. Age the mouthwash for a sufficient length of time to permit all insoluble materials to be precipitated. Remove the supernatant liquid after aging and this should be refrigerated at a temperature of 40 C and then filtered while cold.

No. 4

(Antibiotic)

Gramicidin	0.006
"Pluronic" F-68	0.500
Spearmint Oil	0.050
Menthol	0.050
Cinnamon Oil	0.010
Dye	0.001
Sodium Saccharin	0.150
Alcohol (Ethanol 95%)	25.000
Water	74.233

Add ingredients to water in order shown mixing gently. When system is homogeneous, transfer to suitable containers. Product is a clear, fluid liquid.

No. 5

"Hyamine" 10-X		0.03
"Hamposyl" L-30		0.06
"Pluronic" F-68		0.50
Glycerin		10.0
Ethanol		5.0
Water, Flavor, Color	q.s.	100

Breath Freshener

	Per Tablet
Wintergreen Oil	0.60 mg
Menthol	0.85 mg
Peppermint Oil	0.30 mg
"Syloid®" 244	1.00 mg
Sodium Saccharin	0.30 mg
Sodium Bicarbonate	14.00 mg
Mannitol USP (granular)	180.95 mg
Calcium Stearate	2.00 mg

Mix the flavor oils and menthol till liquid. Adsorb onto the "Syloid®" 244. Add the remaining ingredients. Blend and compress on 5/16 in. flat face bevel-edge punch to a thickness of 3.1 mm.

Toothpaste
Formula No. 1

"Poloxamer" 407	22.0
Glycerin	8.8
Water-Insoluble Polishing Agents	4.0
Sodium Citrate	1.5
Citric Acid	0.5
Flavor	0.7
Coloring Agent	0.0006
Deionized Water	62.4994

No. 2

"Pluronic" F-87	2.0
Water	27.0
Sorbitol	12.0
Glycerin	12.0
Dicalcium Phosphate Dihydrate	43.5
"Cab-O-Sil"	1.5
CMC (95% food grade)	0.5
Sodium Saccharin	0.5
Sodium Benzoate	0.5
Peppermint Oil	0.3
Anise Oil	0.2

Dissolve "Pluronic" F-87 polyol in water. Add sorbitol and glycerin and mix. Add the dicalcium phosphate, silica and cellulose gum previously blended, and mix well until free of lumps. Mix in saccharin and sodium benzoate. Add peppermint and anise oil and mix until homogeneous. Transfer to suitable containers. Product is a liquid which turns into a tube-able paste on aging for 24 h.

No. 3

(Liquid)

A	"Veegum"	1.00
	Cellulose Gum, CMC-7MF	.25
	Distilled Water	21.25
B	Sorbitol (70% sol'n.)	12.50
	Glycerin	12.50
C	Dicalcium Phosphate Dihydrate	50.00
D	Flavor	1.00
E	Sodium Lauryl Sulfate	1.50
	Preservative	as needed

Dry-blend the "Veegum" and cellulose gum. Add to water slowly, agitating until smooth. Mix ingredients in B. Add B and C alternately to A with mixing. Add D and mix until smooth. Mix ingredients in E. Add E with slow mixing, avoiding incorporation of air.

No. 4

Cellulose Gum, CMC-7MF	1.0
Glycerin	24.0
Distilled Water	21.0
Methyl Parasept	0.5
Flavor	as needed
Sweetener	0.2
Sodium Alkyl Sulfate	2.3
Calcium Carbonate	15.0
Dicalcium Phosphate	36.0

Mix cellulose gum with the glycerin of the humectant. Add water to this cellulose gum mix. Heat to 70–80 C and hold for 10 min while stirring. Cool the entire mixture to room temperature and transfer to a suitable mixer. Add the sweetener solution, flavor, and preservative while mixing. Add the synthetic detergent with slow mixing to avoid excessive foaming. Add the abrasive slowly, allowing it to disperse well. After the abrasive is added, mix until the paste is homogeneous.

No. 5

(For Electric Tooth Brush)

A	"Veegum	1.50
	Cellulose Gum, CMC-7MF	.80
	Distilled Water	28.05
B	Sweetener	0.15
	Distilled Water	2.00
C	Sorbitol (70% sol'n.)	15.00
	Glycerin	10.00
D	Dicalcium Phosphate Dihydrate	30.00
	Tricalcium Phosphate	10.00
E	Flavor	1.00
F	Sodium Lauryl Sulfate	1.50
	Preservative	as needed

Dry blend the "Veegum" and the cellulose gum. Add to the water slowly, agitating continuously until smooth. Mix ingredients in B and add to A. Mix ingredients in C and D separately. Add C and D alternately to A and B while mixing. Add E to the mixture and mix until smooth. Mix ingredients in F. Add F to the other components with a minimum of slow mixing.

NO. 6

A	"Veegum"	1.0
	Water	18.5
B	Sodium Carboxymethylcellulose (high visc.)	0.5
	Glycerin	30.0
C	Dicalcium Phosphate	47.0
D	Flavor	1.0
E	Sodium Lauryl Sulfate	2.0
	Preservative	q.s.

Add the "Veegum" to the water slowly, agitating continually until smooth. Wet the CMC with glycerin and add to A slowly with agitation. Add C slowly with agitation. Add D and mix. Add E slowly with just enough mixing to obtain a smooth batch.

NO. 7

A	"Veegum" F	1.25
	Sodium Carboxymethylcellulose (med. visc.)	0.70
	Water	23.40
B	Saccharin	0.15
	Water	2.00
C	Sorbitol (70% sol'n.)	12.50
	Glycerin	12.50
D	Dicalcium Phosphate (dihydrate)	45.00
E	Flavor	1.00
F	Sodium Lauryl Sulfate	1.50
	Preservative	q.s.

Dry blend the "Veegum" F and the CMC. Add to the water slowly, agitating continually until smooth. Add B to A. Add C and D alternately to A and B with mixing. Add E to the mixture, mixing until smooth. Add F to the other components with a minimum of slow mixing to avoid incorporation of air.

Note: This formula will have a thin mixing viscosity but, due to the thixotropic nature of "Veegum," will set up in the tube.

Denture Cleaner

FORMULA NO. 1

A	"Veegum"	1.0
	Sodium Carboxymethylcellulose (med. visc.)	0.5
	Water	28.9
B	Saccharin	0.1
	Sodium Benzoate	1.0
C	Sorbitol (70% sol'n.)	9.0
	Glycerin	9.0
D	Dicalcium Phosphate (dihydrate)	36.0
	Dicalcium Phosphate (anhydrous)	12.0
E	Flavor	0.5
F	Sodium Lauryl Sulfate	2.0

Dry blend the "Veegum" and the CMC. Add to the water slowly, agitating continually until smooth. Add B to A and mix. Blend C and add to A and B. Blend D and add to the mixture. Add E and F one at a time, to other components and mix until uniform.

NO. 2

A	"Veegum" WG	5
	Sodium Perborate	13
	Tetrasodium Pyrophosphate (anhydrous)	25
	Sodium Chloride	13
	Tartaric Acid	9

	Sodium Phosphate (dibasic)	12
	Citric Acid	7
	Sodium Bicarbonate	16
B	Isopropyl Alcohol	25
C	"Veegum" WG	2

Granulate A with B. Pass through a 10-mesh screen. Dry the granulation 1 h at 105 C and pass through a 16-mesh screen. Dry blend with C and compress.

Directions for use: Add one tablet to a glass of hot water. Soak dentures in solution ½ hour or overnight and rinse.

	NO. 3	NO. 4
	(Powder)	*(Cream)*
Cellulose Gum, CMC-7H3SXF	66.0	33.0
Talcum USP	33.0	16.0
Perfume and Color	1.0	1.0
Petrolatum	—	50.0

NO. 5

A	"Vinac" B-15	46
	Ethyl Alcohol	32
B	"Veegum"	1
	Water	18
C	"Atmos" 300	3

Add the "Vinac" B-15 slowly with stirring to the alcohol which has been heated to 65–70 C. Maintain temperature and stir until all the "Vinac" has dissolved and a clear gel is formed. (Maintain the alcohol level by replacing any alcohol that has boiled off.) Add the "Veegum" to the water slowly, agitating continually until smooth. Add B to A with agitation. Stir until smooth. Add C and stir until uniform.

Chapter VII

ELASTOMERS, PLASTICS AND RESINS

Automotive Cellular Extrusions

FORMULA	NO. 1	NO. 2	NO. 3
"Vistalon" 6505	100	100	100
FEF Black	20	—	50
SRF Black	—	55	—
MT Black	50	—	—
Whiting	60	60	75
"McNamee" Clay	60	60	—
Paraffinic Petroleum Oil (43 SSU at 210 F)	50	—	—
"Circolite" 4240	—	55	60
Zinc Oxide	5	5	5
Stearic Acid	2	2	2
"Factice" 57-S	5	5	5
"Kempore" 200	7	7	7
Sulfur	2	2	2
"Vocol" S	2.5	2.5	2.5
"A-1"	3	3	3
ZDMDC	3	3	3
TDEDC (80%)	1	1	1

Microcellular Rubber

"Vistalon" 5600	100
"Mistron" Vapor Talc	160
"Suprex" Clay	200

"Sunpar" 2280	130
Paraffin Wax	5
Zinc Oxide	5
Stearic Acid	3
Diethylene Glycol	2
"Unicel" ND	10
B-I-K	2.5
Sulfur	1.5
TMTDS	1.5
ZMBT	1.5

EPDM Closed Cell Sponge

"Vistalon" 6505	100
Zinc Oxide	5
Stearic Acid	1
MT Black	100
FEF Black	50
"Sunpar" 2280	75
Petrolatum	5
"Celogen" AZ	10
B-I-K	1
Sulfur	2
"Pennac" NB Ultra	2

Crushed Foam Compound

FORMULA NO. 1

"Geon" 576	179.0
"Kronitex" 100	25.0
"Nopco" KOY	16.7
"Titanox" RA-50	33.3
(slurry in water)	
Tap Water	40.0
"Catalpo" Clay	30.0
"Sipex" SB 8208	3.3
Ammonium Stearate	18.0

No. 2

"Geon" 460X2	202.00
Ammonium Stearate	21.23
"Sipex" SB	6.70
Tap Water	78.00
Tetrasodium Pyrophosphate	0.75
"Titanox" RA-50	15.00
"Nytal" 300	130.00
"Good-Rite" K-718	14.30

Molded Open Cell Sponge, Automotive Gaskets

FORMULA	No. 1	No. 2
"Vistalon" 4608	100	100
MT Black	150	90
SRF Black	—	60
Naphthenic Petroleum Oil (84 SSU at 210 F)	90	100
"Unicel" S	15	15
Zinc Oxide	5	5
Stearic Acid	5	5
Sulfur	1.5	1.5
TMTDS	1.5	1.5
MBT	0.5	0.5

Natural Rubber Foam

FORMULA No. 1

A	NR Latex Concentrate	167.0
	Castor Oil Soap Solution	1.0
	Antioxidant Dispersion	2.0
	Accelerator Dispersion (ZDEC)	2.0
	Accelerator Dispersion (ZMBT)	2.0
	Sulfur Dispersion	5.0
	Color Dispersion	2.0
	Filler Slurry	as required

B	Polypropylene Glycol 750 Sol'n.	10.0
	"Vulcastab"	1.0
	Zinc Oxide Sol'n.	6.0

The base mix A is foamed to the required volume. Mixture B is added and the rate of agitation is reduced to refine the foam in the usual manner. The foam is poured into a mold which has been preheated to 55-65 C and the mold is closed. After 5-10 min gelation will be complete and the mold may be heated in steam or hot water at 100 C for approximately 30 min to achieve vulcanization. After vulcanization the product is stripped, washed, and dried.

No. 2

A	NR Latex Concentrate	167.0
	Castor Oil Soap Sol'n.	6.0
	Antioxidant Dispersion	2.0
	Casein Sol'n.	10.0
	Accelerator Dispersion (ZDEC)	2.0
	Accelerator Dispersion (ZMBT)	0.6-2.0
	Sulfur Dispersion	5.0
	China Clay	70-100
	Color Dispersion	2.0
B	Polypropylene Glycol 750 Sol'n.	4.0
	"Vulcastab"	6.0
	Zinc Oxide Dispersion	6.0

The base mix A is placed in the bowl and foamed to approximately six times the original volume of latex. Mixture B is added and the rate of foaming is reduced to refine the foam. At this viscosity the foam is suitable for spreading operations, and thicknesses of up to 5 mm are readily gelled. Gelation is preferably carried out in a hot air stream at 90 C. Drying and vulcanizing may then be carried out at any convenient temperature.

SBR Latex Foam

SBR Latex (65-70%)	292.0
Filler (calcium carbonate clays)	270.0
Hydrated Alumina	50.0

Product E-413 (35%)	11.4
Sodium Lauryl Sulfate (30%)	10.0
Zinc Diethyl Dithiocarbamate	3.0
Sodium Hexametaphosphate (5% sol'n.)	20.0
Zinc MBT	2.0
Water	30.0
Sulfur-Based Vulcanizer	24.0

Foaming Procedure (No-Gel Type)

The latex and water are mixed together with a Lightning Mixer, and the pH adjusted to 10.7 with ammonium hydroxide solution. Product E-413 is added, and the pH readjusted if necessary to 10.7. Sodium hexametaphosphate solution is added slowly to the mixture. Filler, hydrated alumina, zinc diethyl dithiocarbamate, and zinc MBT are premixed and this mixture is slowly sifted into the liquid latex mixture. Mixing is continued for about 15 min to insure uniformity, whereupon the sodium lauryl sulfate and the vulcanizer compound are added and mixing is continued for about 5 min.

The compounded latex is now transferred to a Hobart mixer so that air may be whipped into the latex to produce foam. A Hobart Model 4-C can be used for laboratory scale batches. Speed settings for 2 min each at 6, 4, and 2 positions (in that order) produce the required whipping action. The foamed compounded latex is now quickly applied to the back of carpet swatches by doctoring with a 1/8-1/4 in. guide. The swatch is then given a 30-s IR lamp treatment ($\sim$ 15 in. away) to produce a "skin" on the foam, and then cured in a forced draft oven at 275 F for 15 min.

Foaming Procedure (Gel Type)

If a foamed latex is required with even greater collapse resistance, Product E-413 can be utilized in a gel type foamed latex. The gel type of foamed latex produces an internal gel in the latex compound to be doctored onto the carpet swatch. This imparts greater foam stability duing the curing cycle.

Flexible Polyurethane Foam, Flame Retardant

	FORMULA No. 1	No. 2	No. 3
"Antiblaze" 19	—	5	10
TDI (80/20)	41.8	41.8	41.8
Polyether Triol (3000 MW)	100.0	100.0	100.0
Water	3.2	3.2	3.2
"Y-6634"	1.0	1.0	1.0
"Dabco" 33-LV	0.3	0.3	0.3
Stannous Octoate	0.3	0.3	0.3
R-11	3.0	4.0	5.0

Formulation Information:

NCO/OH Index	1.05	1.05	1.05
% Mobil Antiblaze 19	0	3.3	6.4
% Phosphorus	0	0.69	1.34

Foaming Rate (Hand-Mix Foams):

Stir Time (s)	13	9	9
Rise Time (s)	98	165	153

Rigid Foam Profile

	Single-Screw
PVC Resin (K-68)	100 PHR
Tin Stabilizer	.7
Calcium Stearate	1.0
"Wax XL" 355	.2-.5
TiO_2	1-2
$CaCO_3$	3-5
"Experimental Resin" (CPE) XO-2243.56	4-6
"Celogen" AZRV	.5-.7
"Acryloid" K120N	1-2

Molding Compound

"FaRez" B-260	80
"Polycat" 200	8
OCF 1/8 in. 832 Fiberglass	70
Carbon Flour	90

Toy Balloons

	Prevul-canized latex	Vulcaniz-able mix
Natural Rubber Latex (60%)	167.0	167.0
Potassium Hydroxide Sol'n. (10%)	2.0	4.0
Potassium Caprylate Sol'n. (20%)		2.0
Sulfur Dispersion (50%)		1.0
Zinc Diethyldithiocarbamate Dispersion (50%)		1.5
Antioxidant Dispersion (50%)	2.0	2.0
Zinc Oxide Dispersion (50%)		0.5
Water	3.0	
Color Dispersion	as required	as required
Cure/drying in hot air at 120 C, min	20-25	20-25

Manufacture

Compounding ingredients   The ingredients are used in the form of solutions, emulsions, or dispersions. When preparing dispersions, it is essential that the particle size of the dispersed material is not greater than 5μm. Coarse particles can make processing difficult by settling in the dipping tanks and may also cause defects in the product.

Mixing and maturation   Ingredients are added to the latex in the order given in the formulations and the mix is thoroughly stirred to ensure homogeneity. The vulcanizable mix needs to be "matured" prior to use and should be kept at a temperature of 30 ± 2 C for 16 h with gentle stirring, and then cooled to 15-20 C.

After preparation the mix should be sieved through an 80-100 mesh nylon or stainless steel gauze before being fed to the dipping tank.

Dipping and curing   Porcelain, glass, or aluminum formers are dipped into a 15-20% solution of calcium nitrate in alcohol, to which a small amount (1-5%) of detackifying agent (e.g., talc) has been added. It is advisable to dry the coagulant on the formers to prevent coagulant dripping into the latex bath. The formers are then carefully dipped into the mix and, after the desired dwelling time, withdrawn and dried in an oven at low temperature (60-70 C). The balloons are beaded and then dried and vulcanized in hot-air circulating ovens.

Stripping and drying   Balloons are stripped from the formers either manually or with various automatic stripping devices. Before stripping, the balloons are sprayed with water or with a suspension of fillers of fine particle size in water to facilitate the process. In manually operated plants, dipping rather than spraying of water may be preferred. The stripped, wet balloons are then dried in either static hot-air ovens or cyclone driers.

SBR, Light Color

FORMULA NO. 1

SBR 1502	100
"HiSil" 233	30
Stearic Acid	1
Cumar MH 2½	10
Diethylene Glycol	2.5
"Titanox"	10
ZnO	5
Sulfur	1.75
DPG	1.5
"Zemite"	.75
"Factice" Amberex B	30
"Circo" Light Oil	30

Cure: 15 min at 292 F.

No. 2

SBR 1006	100
"HiSil" 233	40
Stearic Acid	1
Cumar MH 2½	10
Diethylene Glycol	2.5
"Titanox"	15
ZnO	5
Sulfur	1.75
"Zenite"	.75
DPG	1.5
"Factice" Amberex B	15

Cure: 12 min at 292 F.

Rubber Bands

"Natsyn" 200	100
ZnO	5
Stearic Acid	2
"HiSil" 233	5
"Antioxydant 425"	1
Methyl "Tuads"	.1
Altax	1
Sulfur	2
"Factice" Amberex S	20

Cure: 8 min at 293 F.

Inexpensive Isoprene Formula

Isoprene	50
SBR 1805	106
"Neozone" D	1
ZnO	5
Stearic Acid	3

"Philblack" O 50
Sulfur 2
"Santocure" 1

Cure: 30 min at 292 F.

Polybutene Emulsions

FORMULA	NO. 1	NO. 2	NO. 3	NO. 4
Emulsifier Type	*(Anionic)*	*(Nonionic)*	*(Nonionic)*	*(Cationic)*
Polybutene	100	100	100	100
Oleic Acid	3.5	–	–	–
Triethanolamine	1.7	–	–	–
"Triton" X-100	–	4	–	–
"Siponic" 218	–	–	5.0	–
Isothan DL-1	–	–	–	5.0
Water	67.3	29.3	67.3	67.3

Polyvinyl Acetate Latex

Potassium Persulfate (1% sol'n.)	30
"Aerosol" A-103 (10% sol'n.)	30
Sodium Bicarbonate (1% sol'n.)	10
tert-Dodecyl Mercaptan	0.18
Vinyl Acetate (polymerization grade)	100
Boiled Deionized Water to Give 40% Polymer Solids	80

Equipment:

Three-necked, 250 ml round bottom flask, mantle, stirrer, condenser, thermometer, adapters, addition funnels, and tank of prepurified nitrogen.

Prepare fresh solutions of potassium persulfate, surfactant and sodium bicarbonate using boiled, deionized water. Slowly bubble nitrogen through the potassium persulfate solution for 10 min to expel oxygen and set aside in an addition funnel.

Charge the prescribed amounts of surfactant and sodium bicarbonate solutions and boiled, deionized water to the reaction vessel. Slowly bubble nitrogen under the surface of the liquid for about 15 min. Add the *tert*-dodecyl mercaptan and adjust the nitrogen inlet tube just above the surface of the liquid. Adjust the stirrer speed to about 200 rpm. Heat the reaction mixture to 65 C. Then, add 3 ml of the potassium persulfate solution.

Continue heating for 10 min. Begin the continuous dropwise addition of monomer and potassium persulfate solution. Adjust the addition rates so that, after 2 h, all of the monomer is added and about 3 ml of the potassium persulfate solution remains. During the addition period, maintain gentle refluxing at temperatures between 67 and 73 C. Then, add the remaining potassium persulfate solution over a 15-min period. Continue the polymerization for one additional hour to insure complete conversion of monomer. Cool and filter the latex to determine the coagulum content and set aside for later evaluation.

Copolymer Latex

Vinyl Acetate (H grade)	255
2-Ethylhexyl Acrylate	45
$K_2S_2O_8$ (fresh 1.5% sol'n.)	60
$NaHSO_3$ (fresh 1% sol'n.)	30
$NaHCO_3$	0.3
"Aerosol" A-102 Surfactant (35% sol'n.)	25.7
Boiled Deionized Water	58

Equipment:

1 liter, 3-necked, round bottom flask, mantle, stirrer, condenser, thermometer, adapters, addition funnels and tank of prepurified nitrogen.

Prepare fresh solutions of $K_2S_2O_8$ (1.5%) and $NaHSO_3$ (1%) using boiled, deionized water and purge for ca. 10 min with a stream of prepurified nitrogen. Charge to the reaction vessel all of the required amount of surfactant solution, all of the bicarbonate and the boiled deionized water. Purge under the surface of the liquid with a stream of prepurified nitrogen for ca. 15 min. Add 3 ml of the bisulfite solution, 15 ml of the persulfate solution, and 30 g (~33 ml) of the monomer mixture. Adjust the gas inlet tube to purge the air space above the liquid. Set the stirrer speed at ~250 rpm and begin to heat to reflux temperature. After the polymerization exotherm subsides (initial monomer charge polymerized), begin the continuous addition of the remaining monomer mixture, persulfate solution and bisulfite solution. Adjust addition rates so after 3 h all of the monomer mixture is depleted and only a few ml of each of the initiator solutions remain. During the addition, maintain a gentle reflux (temperature of 65-70 C). After additions are complete, continue polymerization for another hour. Cool and bottle for subsequent evaluations.

Acrylic Latex

Kettle Charge:

"Aerosol" A-103 (35% sol'n.)	11.4
	(2% real)
Potassium Persulfate (5% sol'n.)	20
Methanol	5
Water	92

Preemulsified Monomers:

Ethyl Acrylate	134
Methyl Methacrylate	60
Sodium Metabisulfite	0.4
Water	65.5
Methanol	5
"Aerosol" MA-80 Surfactant	2.5
	(1% real)

Delayed Portion:

Itaconic Acid	2.0
N-Methylolacrylamide (60%)	6.7
Diammonium Phosphate	0.3
Water	10.0

Preparation of Kettle Charge

Add 11.4 parts of "Aerosol" A-103 surfactant and 5 parts of methanol to 92 parts of boiled, nitrogen purged water. Dissolve 1 part of potassium persulfate in 19 parts of water, add to the above solution and purge with nitrogen for 15 min. After purging, place solution in polymerization kettle and stir while heating to 60 C. Continue to purge with nitrogen while heating.

Natural Rubber Latex Stabilizer

A	Caprylic Acid	14.5
	Soft Water	40.0
B	Potassium Hydroxide	5.8
	Soft Water	40.0

A—warm the soft water to 75 C then add caprylic acid with fast (400–600 rev/min) stirring.

B—dissolve potassium hydroxide in water.

Slowly add B to A with stirring. Allow to cool before use.

Latex Casting

	FORMULA NO. 1	NO. 2	NO. 3
	(Soft)	*(Semi-rigid)*	*(Rigid)*
NR Latex (60%)	167.0	167.0	167.0
Surfactant (10%)	–	–	5.0
Sulfur Dispersion (50%)	2.0	3.0	5.0
Zinc Oxide Dispersion (50%)	2.0	3.0	10.0
Zinc Diethyldithiocarbamate			
Dispersion (50%)	2.0	2.0	3.0
Whiting	–	100.0	300.0
Na₄ Pyrophosphate	–	0.5	1.0
Water	–	30.0	90.0
Cure time/temperature,			
min/°C	30/100	30/100	30/100

Casting

Sieve the latex mix into the mold, taking care not to entrap air. Allow the filled mold to stand for the time necessary to build up the required deposit on the mold surface. (Topping up may be necessary during the standing period.) Pour the excess latex from the mold and allow the mold to drain for a few minutes. Heat the mold (up to 90 C) to dry the product sufficiently to enable it to be removed without distortion. (NB–A greater degree of drying is necessary with unfilled mixes.) Heat the casting at 100 C in hot air to effect final drying and/or vulcanization.

The vulcanized casting is then ready for finishing.

Finishing

Where articles of one base color are required, coloring agents can be incorporated into the filler slurry before mixing. Where more than one color is desired surface finishes are applied by spray or brush. These finishes should be as flexible as the rubber itself. Plasticized cellulose lacquers are suitable for rigid castings, but more flexible finishes are required for softer products, and paints based on polychloroprene or polyurethanes are suggested.

With high filler contents, castings are sufficiently hard for seam marks to be removed by sandpaper, knife or buffing wheel.

Flameproof Plastics

FORMULA NO. 1

(High Impact Polystyrene)		Oxygen Index	1/8 in. Bar UL-94
Add			
*DBBPE	10-13%	23	V-O
Antimony Trioxide	3-6%		

NO. 2

(Low Density Polyethylene)			
*DBBPE	6%	24	V-2
Antimony Trioxide	2%		

No. 3

(Thermoplastic Polyester)

*DBBPE	6–8%	—	V-O
Antimony Trioxide	2–3%		

*Decabromodiphenyl Ether

No. 4

(Smoke Depressant)

Polyvinyl Chloride Resin	100.0
2-Ethylhexylphthalate	7.0
Epoxidized Soybean Oil	3.5
Stearic Acid	4.0
Ba/Cd/Zn Stabilizer	0.5

No. 5

Polyvinyl Chloride Resin	100.0
2-Ethylhexylphthalate	50.0
Epoxidized Soybean Oil	7.5
Stearic Acid	8.3
Ba/Cd/Zn Stabilizer	0.8

No. 6

Polyethylene	100
Chlorinated Paraffin	20
Trioxide	5

This is a white opaque material. LOI = 25.9; in horizontal burning, S.E. in 10 s, burning drips.

No. 7

Polyethylene	100
Chlorinated Paraffin	20

This is a gray translucent material. LOI was 21.9; continued burning, burning drips.

No. 8

Polyethylene	100
Chlorinated Paraffin	20
"Nyacol" A1588LP	5

This is a light tan, translucent material. LOI = 27.2; S.E. in 9 s, drips, 1 drip burnt.

No. 9

Polyethylene	100
Chlorinated Paraffin	20
"Nyacol" A1588LP	3

This is a light tan, translucent material. LOI = 27.6; S.E. in 8 s, drips, 2 drips burnt.

No. 10

Polyethylene	100
Chlorinated Paraffin	20
"Nyacol" A1588LP	1.5

This is a very light tan, translucent material. LOI = 26.8; S.E. in 8 s, drips, no burning drips.

Pressure Pipe Plastics

	(Twin-Screw)	*(Single-Screw)*
PVC Resin	100 PHR	100 PHR
Tin Stabilizer	.3-.4	.4-.8
Calcium Stearate	.5-.8	.6-1.0
"Wax XL" 355	.8-1.2	.6-1.0

TiO$_2$	1–2	1–2
CaCO$_3$	3–5	3–5
"Experimental Resin" (CPE)		
XO-2243.56	2	2

UL Conduit Plastics

	(Twin-Screw)	*(Single-Screw)*
PVC Resin (K-68)	100 PHR	100 PHR
Tin Stabilizer	.3–.4	.4–.8
Calcium Stearate	.5–.8	.6–1.0
"Wax XL" 355	.8–1.2	.6–1.0
TiO$_2$	1–2	1–2
CaCO$_3$ Filler	3–5	3–5
"Experimental Resin" (CPE)		
XO-2243.56	4–6	3–5

Sewer Pipe Plastic

	(Twin-Screw)	*(Single-Screw)*
PVC Resin (K-68)	100 PHR	100 PHR
Tin Stabilizer	.3–.4	.4–.8
Calcium Stearate	.5–.8	.6–1.0
"Wax XL" 355	.8–1.2	.6–1.0
TiO$_2$	1–2	1–2
CaCO$_3$ Filler	25–40	25–40
"Experimental Resin" (CPE)		
XO-2243.56	4–6	3–5

Vinyl House Siding

	(Twin-Screw)	*(Single-Screw)*
PVC Resin (K-68)	100 PHR	100 PHR
Tin Stabilizer	.5	.7
Calcium Stearate	.7	1.0
"Wax XL" 355	1.0	1.0

TiO$_2$	10–13	10–13
CaCO$_3$	3	3
"Experimental Resin" (CPE)		
XO-2243.56	4–6	6–8
"Acryloid" K120N	1–2	.5–1.5

Vinyl Plastisol

FORMULA NO. 1

"Geon®" 121	100.0
DOP	43.0
"Santicizer" 160	15.0
Dry "Dythal"	2.0
"Carbopol" 940 Resin	1.0
"Ethomeen" C-25	3.0

Put dry "Geon®" 121 resin, "Carbopol" 940 resin, and dry "Dythal" in a Hobart bucket. Use only "Dythal" powder, not paste. Start mixing and add blend of plasticizers. After plastisol is made and let down with remaining plasticizer, then deaerate. To the deaerated plastisol, stir in the "Ethomeen" C-25 and wait for complete viscosity development—approximately 1 h.

NO. 2

"Geon®" 440X24	20
MEK	40
"Cellosolve"	60
"Carbopol" 941	3

Resin Emulsion

"Zonester" 85 Resin Ester	40
"Acintol" FA-1 Fatty Acid	0.7
Mineral Spirits	10
KOH Sol'n. (3%)	6
Water	35

Dissolve the "Zonester" 85 and FA-1 in the mineral spirits. To this solution add the aqueous KOH with stirring to form a homogeneous dispersion. Then add the water to adjust the solids to 45%. This procedure results in an emulsion of moderate stability. If an emulsion with better storage stability is required, then the use of a high shear mixer is beneficial. Again, if better shelf life is required, 15 parts of a 10% ammonium caseinate solution may be used with the corresponding reduction in water.

Cleaning Empty IPDI and TMDI Containers

Empty IPDI and TMDI containers are cleaned like those for any other isocyanate. All isocyanate residues are substantially removed and replaced with about 2-3 liters of a solution of

Methanol	30
Water	70
Ammonia (conc.)	1

Roll and turn the containers to make sure that all surfaces are wetted. The drum may be emptied after allowing the solution to react for at least 12 h (overnight). Rinse repeatedly with water.

Removing Spilt IPDI or TMDI

Any isocyanate spilt from leaky containers or through carelessness must be soaked up at once with dry sawdust. Pour dilute ammonia over it, and then keep it in closed containers until you are ready to destroy the isocyanate by burying it in damp soil. Do not burn the sawdust under any circumstances. Rather than absorbing it in sawdust, it is better to sprinkle any isocyanate pools with an effective powder of the following composition:

Carrier:	
Sawdust	23.0
Kieselguhr	38.5

Destructive solution:

Denatured Alcohol	19.4
Triethanolamine	3.8
Ammonia (conc.)	3.8
Water	11.5

(about 42.5 l)

The sprinkling powder is prepared in an efficient mixing drum. It is sprinkled as a thin layer on the isocyanate residues (about four times as much). In this way the isocyanate is absorbed and destroyed within a short time. The isocyanate-soaked sprinkling powder is also eliminated by burying it in the soil.

Instead of the above sprinkling powder, a spray mix of the following composition may be used to destroy isocyanate residues:

Ethanol	50
Water	40
Ammonia (conc.)	1–10

Solvents

FORMULA NO. 1

(For Polyurethane)

"NiPar" S-30 or "NiPar" S-20	19
Dimethylformamide	68.5
Isopropyl Alcohol (99%)	12.5

NO. 2

(For Polyurethane)

"NiPar" S-30 or "NiPar" S-20	24
Hexane	21
Dimethylformamide	55

No. 3

(For Phenoxy Coating)

"NiPar" S-30 or "NiPar" S-20	20
Toluene	19
Acetone	30
Isobutyl Acetate	11
"Shellacol"	20

No. 4

"NiPar" S-30	24
Toluene	19
Isobutyl Alcohol	32
Methyl Ethyl Ketone	25

No. 5

(For Polystyrene)

"NiPar" S-30 or "NiPar" S-20	20
Heptane	60
Toluene	20

No. 6

(For Thermoplastic Acrylics)

"NiPar" S-30	15
Ethylbenzene	16
Methyl Ethyl Ketone	25
Heptane	30
Cyclohexanone	14

No. 7

(For Thermosetting Acrylic)

"NiPar" S-30	17.5
Hi-Flash VM&P Naphtha	51.0
Methyl Amyl Acetate	20.0
"Cellosolve" Acetate	7.5
Xylene	4.0

No. 8

(For Acrylic-Nitrocellulose Inks)

"NiPar" S-30	12
Isopropyl Acetate	38
"Shellacol" (anhydrous)	50

No. 9

(For Short Oil Alkyd)

"NiPar" S-30	24.3
Naphtha	47.8
"Cellosolve" Acetate	8.7
Toluene	14.8
Isobutyl Alcohol	4.4

No. 10

(For Ethyl Cellulose)

"NiPar" S-30	5
Toluene	20
"Shellacol"	35
Heptane	40

No. 11

(For Ethylhydroxyethyl Cellulose)

"NiPar" S-30	7.5
Heptane	84.0
Isopropyl Alcohol (99%)	8.5

No. 12

(For Flexographic Ink)

"NiPar" S-30 or "NiPar" S-20	10
"Shellacol"	70–80
Ethyl Acetate	10–20

	No. 13	No. 14
	(For Cellulose Propionate)	
"NiPar" S-30	10	10
"Shellacol"	90	71
Isopropyl Acetate	0	19

No. 15
(For Nitrocellulose)

"NiPar" S-30	10–15
Toluene	20
Isopropyl Alcohol (99%)	17
Heptane	25–28
Isobutyl Acetate	20–25
Methyl Ethyl Ketone	5

No. 16
(For Chlorinated Rubber)

"NiPar" S-30	15
Methyl Ethyl Ketone	55
Heptane	30

No. 17
(For Epoxy Resins)

"NiPar" S-30	11.0
Toluene	7.0
Methyl Ethyl Ketone	10.5
Ethylene Glycol Monoethyl Ether Acetate	3.5
Cyclohexane	3.0

No. 18
(For Vinyl)

"NiPar" S-30	30
Cyclohexanone	20
Aromatic Solvent	8
Ethylbenzene	11
Naptol	31

No. 19

(For Polyamides)

"NiPar" S-30	10–25
"Shellacol"	80–65
"Lactol" Spirits	10

Thickened Epoxy Systems

FORMULA NO. 1

DMF	45
MEK	45
DI-2 Ethyl Hexyl Amine	7
"Carbopol" 934	3
"DER" 736 or "Epon" 828	25
MeOH	10

No. 2

"Epon" 1007	25
Dicyandiamide	2
"Carbitol"	76
"Carbopol" 941	2

Castable Epoxies

	FORMULA NO. 1	NO. 2
"Hycar" ATBN (1300 × 16)	300	300
DGEBA Liquid Epoxy Resin	100	—
Bisphenol A	24	—
"ERL"-4221	—	100
Hexahydrophthalic Anhydride	—	100
Cure Cycle: T/t-°C/min	120/18.5	150/3

Chapter VIII

FOOD PRODUCTS

Synthetic Beverages

Synthetic beverages can be made from the following edible materials:

1. A carrier, e.g., sugar, lactose, dry milk, etc.
2. A flavor, e.g., coffee flavor.
3. An energizer, e.g., caffeine, niacin.
4. A non-caking powder, e.g., fine sodium silico aluminate.
5. Brown food color.

Other carriers may be used in place of sugar, e.g., soluble starches, dextrose, fructose, lactose, sorbitol, micro-crystalline cellulose, powdered dry milk.

Non-caking alternatives are hydrated silicon dioxide, starch, Avicel.

Synthetic Coffee

Formula No. 1

Coffee Flavor	2–4
Sugar	100
Caffeine	2
Sodium Silico Aluminate	1–2
FDC Food Color	to suit

No. 2

Coffee Flavor	2–4
Lactose	100
Caffeine	2
Hydrated Silica	1–2
FDC Food Color	to suit

No. 3

Coffee Flavor	2–4
Powdered Dry Milk	100
Caffeine Citrate	3–5
Soluble Starch	2–4
FDC Food Color	to suit

Synthetic Cocoa

Synthetic Cocoa Flavor	2–4
Powdered Lactose	100
Caffeine	2
"Avicel"	2
FDC Food Color	to suit

Synthetic Tea

*Natural Tea Extract	5
Fructose	100
Caffeine	2
Sodium Silico Aluminate	1–2

*The tea flavor is a natural extract of tea leaves made by extracting the latter with warm water and concentrating by subjecting it to a vacuum.

Each of the above formulas must be mixed so as to thoroughly disperse the individual ingredients.

Synthetic Cocoa

Synthetic Cocoa Flavor	2-4
Powdered Toasted Carob Flour	100
Caffeine	2
FDC Food Color	to suit

Synthetic Coffee

Coffee Flavor	2-4
Dextrose	100
Niacin	0.4
Hydrated Silica	1-2
FDC Food Color	to suit

Protein Beverage for Developing Countries

Sweet Whey	41
Full Fat Soy Flour	37
Soybean Oil	12
Corn Syrup Solids	9
Fortified with Vitamins and Minerals	1

Natural Nutritious Beverage Powders

Freeze dried fruits, vegetables, coffee, and tea extracts are mixed with "Dry-Mol" (molasses, bran, and starch) and sucrose, fructose, or dextrose. Proteins in the form of soybean, corn or wheat proteins may be included.

FORMULA NO. 1

Freeze Dried Orange Juice	50
"Dry-Mol"	40
Soybean Protein Isolate	10
Dextrose	10

NO. 2

Freeze Dried Mixed Vegetable Juices	40
Dried Skim Milk Powder	10
Sucrose	10

Hot Cocoa Beverage

FORMULA NO. 1

Sugar	45.0
Dextrose	10.0
"Kocoa Kreme" 201, 202 or 203	30.0
Salt	1.0
Cocoa	14.0

NO. 2

Sugar	46.0
"Kocoa Kreme" 203	30.0
Salt	1.0
Cocoa	14.0
Dextrose	9.0

NO. 3

Sugar	47.0
"Kocoa Kreme" 204 and 206	16.0
Dextrose	8.5
Salt	1.0
Cocoa	15.0
Nonfat Dry Milk	12.5

Note: For vending application .5% flow agent to be added.

All formulas should be mixed in the following manner:
1 oz of product to 5½ fluid oz of hot water. Yield: One 6 fluid oz drink.

Orange Milk Drink

Skimmed Milk	42.130
Water	41.575
Sucrose	11.010
Dextrose	3.000
Malic Acid	0.775
20% Sol'n. of a 50–50 Blend of Sodium Benzoate and Potassium Sorbate	0.500
"Genu" Pectin Type JM	0.450
FD&C Yellow No. 6 (1% in sucrose)	0.360
Nat. and Art. Orange Juice Flavor	0.200

Butterscotch Syrup

42 DE Corn Syurp	40.00
Cane or Beet Sugar	31.00
Dark Brown Sugar	2.00
Butter	4.00
Milk-Solids-Nonfat	2.00
Salt	0.10
"NPL"-222	0.50
Flavor Base	0.20
Preservative	0.38
Water	19.83

Chocolate Syrup

42 DE Corn Syrup	25.60
Cane or Beet Sugar	31.00
Cocoa	8.00
Hydrogenated Vegetable Fat	6.00
Milk-Solids-Nonfat	2.00
"NPL"-222	0.25
Preservative	0.38
Water	26.77

Cocoa Butter Replacement

FORMULA NO. 1

(Pastel Coating)

Sugar	46.33
Butterfat Whole Milk Powder (26%)	4.00
Nonfat Dry Milk	15.00
"Satina" II NT	34.00
Salt	.10
Methyl Vanillin	.07
Color (FD&C LAKE)	.10
Lecithin	.40

NO. 2

(Milk Chocolate Coating)

Sugar	42.45
Cocoa Powder	6.00
Butterfat Whole Milk Powder (26%)	8.00
Nonfat Dry Milk	12.00
"Satina" II NT	31.00
Salt	.10
Methyl Vanillin	.05
Lecithin	.40

Clear Natural Anise Flavor

A	Ammoniated Glycyrrhizin	15
	Water	33
	Propylene Glycol	52
B	Oil Anise	16
	Triacetin	56
	Alcohol (95%)	28

Finished Flavor:

A	20
Propylene Glycol USP	78
B	30

Sugarless Chocolate Coating

Cocoa Powder	5.7
Sorbitol, Crystalline	40.3
Nonfat Dry Milk (super heat)	22.0
Hydrogenated Vegetable Fat	31.5
Lecithin	0.5

Sugarless Hard Candy

Sorbitol Crystalline, Tablet Type or Coarse Powder	97.3
Citric Acid (anhydrous)	1.0
Dry Flavor	0.5
Certified Lake Dye	0.2
Magnesium Stearate	1.0

Blend the sorbitol, citric acid, flavor, and dye until uniform. Add the magnesium stearate and blend 10 min. Compress into tablets.

Caramels

FORMULA NO. 1

Water	83
"Richmix" A	40
Sweet Whey Powder	25
Mono- and Diglycerides	1
Granulated Sugar	75
Corn Syrup (42 DE–78% T.S.)	110

Place above ingredients in the above order in a steam jacketed process vessel with agitation. Cook batch to 242 F. Add and incorporate the desired amount of caramel flavor. Pour onto oiled cooling slab and allow to cool.

NO. 2

Sugar	34.66
Corn Syrup	34.66
Fresh Cream (20% fat)	10.83

Powdered Cream	14.44
Dairy Butter	4.33
Salt	0.72
Gelatin	0.36
Vanilla	to taste

No. 3

Sugar	37.98
Corn Syrup	33.23
Creamery Butter	20.18
Nonfat Dry Milk	5.94
Hard Fat	2.37
Salt	0.30
Flavor	to taste

Chocolate Fudge

FORMULA NO. 1

Water	83
"Richmix" A	30
Corn Syrup (43 DE-78% T.S.)	30
Granulated Sugar	120
Brown Sugar	20
Chocolate Liquor	20
Confectionery Hard Fat	10
Fondant (80/20)	70
Salt	to taste

Place above ingredients (except Fondant) in the above order in a steam jacketed process vessel with agitation. Cook batch to 240 F. Add and incorporate desired level of vanilla flavor. Pour into cream beater and circulate cool water to bring temperature to 90 F. Do not mix until cooled. Add required amount of Fondant and proceed to fold. Remove from cream beater and place in oiled pans and allow to cool. To obtain the desired texture, add Fondant after cooking.

No. 2

Sweetened Condensed Whey	43.00
Milk Fat (as butter or cream)	2.50
Sugar	11.00
Corn Syrup	9.00
Invert Syrup	3.00
Fondant	20.00
Chocolate	6.00
Powdered Lactose (for seed)	0.10
Nuts (optional)	5.40
Vanilla	to taste

Raisin and Nut Protective Coating

Zein	27
Stearic Acid	5
Oleic Acid	2
Ethyl Alcohol	63
Water	3

The stearic acid is dissolved in alcohol at 35–40 C. The oleic acid or propylene glycol is added and the solution cooled to 30 C. Then the zein is added slowly with good agitation, and stirred until dissolved. If desired, the formulation may be further diluted with 90% aqueous ethanol.

Eggless Custard

A	Sugar	9.75
	"Redisol" 4	9.50
	Powdered Shortening	2.75
	Salt	1.75
	"Emplex"	.50
B	"Kopald"	8.00
C	Water–Boiling	160.25

A–Combine dry ingredients into uniform mix.
B–Combine with above, blend in a mixer and whip for 10 min or

until desired specific gravity is obtained (approx. 0.75).

C—Meter shortening into blend; mix and pass through finisher if to be stored.

Sugarless Frozen Dessert

Cream (35% fat)	34.3
Sorbitol Sol'n. (70%)	21.4
Water	34.7
Nonfat Milk Solids	9.3
Stabilizer	0.35

Sugarless Jelly

Grape Juice	55.00
Sorbitol Sol'n. (70%)	40.00
Water	3.53
Low Methoxyl Pectin	1.25
Calcium Chloride	0.07
Citric Acid (anhydrous)	0.10
"Sorbistat®"	0.05

Combine the water, low methoxyl pectin, sorbitol solution, and grape juice. Heat to 190 F. Add the calcium chloride, citric acid, and "Sorbistat®." Cool and fill.

Sugarless Chocolate Brownie

A	Unsweetened Chocolate	75.0
	Vegetable Shortening	200.0
B	Whole Egg	140.0
	Vanilla (to taste)	1.0
C	Cake Flour	100.0
	Sorbitol Crystalline (powder)	150.0
	Salt	3.0
	Baking Powder	3.0
	"Veltol®"-Plus (ethyl maltol)	0.01

Melt A. Beat B for 10 min. Add A to B. Mix until smooth. Fold in C. Mix until smooth. Scale 16 oz into 8-in. cake pan. Bake at 350 F for 25 min.

Sugarless Pound Cake

A	Sorbitol Crystalline (powder)	120.0
	Vegetable Shortening	65.0
B	Cake Flour	100.0
	Salt	3.0
	Baking Powder	1.25
	Nonfat Milk Solids	10.0
C	Water	37.5
D	Whole Egg	70.0

Cream A. Add B gradually. Mix. Add C, mix, scrape down. Mix until smooth. Add D, mix, scrape down. Mix until smooth. Scale 24 oz per loaf pan (paper-lined). Bake at 350 F for 65–70 min.

Sugarless Shortbread Cookie

A	Sorbitol Crystalline (powder)	30.0
	Vegetable Shortening	22.5
	Butter	22.5
	Salt	1.25
	Nonfat Milk Solids	5.0
B	Whole Eggs	10.0
	Vanilla (to taste)	1.25
C	Cake Flour	100.0
	Baking Powder	1.25
D	Water	10.0

Cream A until light. Incorporate B. Add C and D. Mix until partially smooth. Roll out on dusted bench or canvas. Cut into various shapes; scale 4 oz per dozen. Place on lightly greased pan. Bake at 390–400 F for 7–9 min.

Sugarless Vanilla Flavored Sandwich Cookie Filling

A	Hydrogenated Vegetable Shortening	42.4
	Lecithin	0.3
B	Sorbitol Crystalline (powder)	28.3
	Salt	0.5
	Vanilla (to taste)	0.2
	"Veltol®"-Plus	0.002
C	Sorbitol Crystalline (powder)	28.3

Melt A. Add B gradually. Mix. Add C. Mix until smooth and uniform.

Doughnut Glaze

A	Sucrose (granulated)	9.25
	Water	11.25
B	Sucrose (granulated) ⎫	2.50
	Salt ⎪ Stabilizer Base	.25
	Vegetable Fat ⎪	.25
	Keylated Agar ⎭	.50
C	Sucrose (powdered)	30.00

A—Blend sugar and water, heat to 160 F.

B—Blend in stabilizer base and maintain temperature at 160 F.

C—Add syrup in 2 phases to powdered sugar, blending until smooth both times. Apply at 135–145 F.

Yeast-Protein Doughnuts

(Winter Wheat)

Flour	100
Defatted Soy Flour	12
Corn Sugar	8
Yeast	6

Salt	2
Shortening	6
"Emplex"	0.5
Nonfat Milk Solids	0.5
Baking Soda	0.5
Sodium Acid Pyrophosphate	0.5
Water	80

Mix 1 min at number one speed; plus 2 min at number two speed in mixer with bowl and dough hook; ferment 1 h, punch dough, and allow dough to rest 20 min; roll dough to yield a 36 g dough piece with a 2½ in. cutter; proof 20 min at 107 F, 85% relative humidity, fry 50 s per side at 385 F.

"Soy Flour" Bread

Ferment:

Flour	100.0
Soy Flour	12.0
Water	78.0
Yeast	3.0
Yeast Food	0.5
Lard	2.0
"Emplex"	0.5

Mix 3 min, number one speed in Hobart A-200 Mixer with McDuffee bowl, at 83 F.

Add at Remix:

Salt	2.5
Corn Sugar	5.0

Mix 2 min at number one speed plus 3 min at number two speed; dough temperature, 83 F; floor time, 45 min; scale 19 oz; 10 min intermediate proof, mold and pan; panary proof to ½ in. above top of pan at 107 F and 85% relative humidity; bake 20 min at 435 F.

Bread Sticks

	lb	oz
Patent Flour	9	8
Gluten		8
Dextrose		3¼
"Unimalt"		3¼
Shortening		13
Yeast		8
Water	5	13
Flavor (optional)		9¾
"Emplex"		1¾

Mix all ingredients at number one speed for 10 min; dough temperature 83 F; 15 min floor time at 83 F and 95% relative humidity. Sheet dough to ¼ in. thickness and cut into ¼ -3/8 in. strips. Cut strips to desired length, usually 4–6 in. long. Lay on flat sheet pan (greased lightly) and proof for 40 min at 107 F and 85% relative humidity. Bake 17 min, with steam, at 380 F.

Sponge Layer Cake

A	Flour	100
	Sugar	120
	Skim Milk Powder	8
	"Vanlite"	5
	Baking Powder	3
	Salt	3
	Water	37
	Whole Eggs	70
B	Water	30.5

A—Mix with wire whip at medium speed.

B—Add gradually. Mix at medium speed until 0.48–0.56 specific gravity is obtained.

Final temperature of the above batter should be approximately 70 F.

Pie Crust

Pastry Flour	100
Shortening	45
Salt	2
Water	25
"Emplex"	1

Add "Emplex" (SSL) with dry ingredients.

Liquid Whipped Topping

Vegetable Fat (101–103 F m.p.)	30.00
Sucrose	6.00
Sodium Caseinate	2.00
Corn Syrup Solids (24 DE)	2.00
Dipotassium Phosphate	.10
CMC (7M)	.45
"Emplex"	.20
Polysorbate 60	.14
Water	q.s. to 100
Flavor and Color may be added as needed	

Melt fat and emulsifiers together at lowest possible temperature. Blend dry ingredients. Add dry ingredients to heated water (120–130 F). Add melted fat and emulsifiers. Pasteurize. Homogenize at 1500 psi and 500 psi (2000 psi), cool rapidly to 40 F and store at 40 F for at least 12 h before whipping.

Frozen Sweet Roll Icing

Water	20.00
Granulated Sugar	25.00
Powdered Shortening	2.00
"Emplex"	.15
Powdered Sugar	100.00

Bring water and granualated sugar to rolling boil. Add SSL and powdered shortening to syrup. Add mixture to powdered sugar slowly and mix until smooth. Apply at a minimum of 120 F. Cool, package, and freeze.

Summer Creme Icing

	lb	oz
Sugar (6x to 12x)	100	
Nonemulsified Shortening	27	
"Vanease"		12
Nonfat Milk Solids	2	8
Salt		9
Water	9	8
Dextrose	10	
Egg Whites	10	
Vanilla Flavor		2

Melt together:

	lb	oz
Nonemulsified Shortening	12	
Vegetable Butter (m.p. 116 F)	6	

After break-up of first stage, add melted mixture of shortening and vegetable butter during mixing. Mix to specific gravity of 0.70.

Creme Filling

	lb	oz
Sugar (6x to 12x)	65	
Nonemulsified Shortening	22	
Salt		8
"Vanease"		11
Vanilla Flavor		5
Butter Flavor		1
Water, Cold	9	8

Add all ingredients to mixing bowl and mix to specific gravity of 0.55–0.59.

Coffee Whitener

	FORMULA NO. 1	NO. 2
Vegetable Fat (102–104 F m.p.)	10	8.5
Corn Syrup Solids (24 DE)	10	6.0
Sodium Caseinate	2.0	0.75
"Emplex"	0.2	0.2
Modified Food Starch	—	0.2
Buffer Salt (dipotassium phosphate)	0.15	0.15
Beta Carotene	0.0004	0.0004
Water	77.6	84.2

Ice Cream Sweetener

	FORMULA NO. 1	NO. 2
"CornSweet" 42 (42% fructose)	25	30
Sucrose	25	30
Corn Syrup (36 DE)	50	40

Water Ice

Cane Sugar	23.0
Corn Syrup Solids	9.0
Stabilizer	0.5
Citric Acid (50%)	0.25
Flavor and Water	67.25

Sherbert

Cane Sugar	22.5
Corn Syrup Solids	8.0
Ice Cream Mix	9.5
Stabilizer	0.5
Citric Acid (50%)	0.5
Flavor and Water	59.0

Low Calorie French Dressing

Water	56.0
Vegetable Oil	12.0
Vinegar (85 grain)	10.0
Salt	3.0
Frozen Egg Yolk	1.0
Sugar	10.0
Spices	3.5
Tomato Paste (21% solids)	4.0
Xanthan	0.15
"Saladizer" #206	0.35

Add egg to water. Add xanthan and "Saladizer" #206 dispersed in oil to above. Let gums hydrate 3-4 min. Add tomato paste, sugar and spices. Add oil. Add vinegar and salt slowly. Pass through colloid mill @ .020-.030 in.

Low Fat Margarine

Aqueous Phase

Water	94
"Supro®" 620 soy isolate	6
Lactic acid	q.s. to pH 5.3

Place water in jacketed kettle. Add protein, stir to disperse, and heat to 90 C or higher. Cool to about 60 C. Add sufficient dilute lactic acid to lower the pH to 5.3.

Oil Phase

Margarine Oils	96.5
Salt	1
"Myverol" 1801	1
Sodium Stearoyl-2-Lactylate	1.5
Color and Flavor	q.s. to taste

Place oils in steam jacketed kettle and heat. Add monoglycerides and lactylate. Melt at 60 C (140 F). Add salt, coloring, and flavoring. Disperse.

Final Process

Add fat phase at 60 C to aqueous phase at 60 C. Disperse with mixer until homogeneous. Pump through normal blending, whipping, and cooling process.

Barbeque Sauce

Water	53.07
Tomato Paste (31% solids)	15.23
Vinegar (100 grain)	13.0
Spices	13.0
Sugar	4.0
Salt	1.0
"Ticaloid" #201R	0.7

Add water. Add tomato paste. Add "Ticaloid" #201R preblended with spices. Add vinegar. Heat to 160 F. Fill.

Nonagglomerating Frozen Foods

Foods containing moisture, when frozen, become hard and must be chopped or thawed before use to get suitable portions and small pieces or units. Such foods bought frozen are fowl, meats, fish, eggs, soy protein mixes, vegetables, fruits, berries, baked goods, dairy products, ice cream, and various mixes.

The addition of pure glycerin to frozen foods prevents agglomeration and the formation of hard masses. The amount of glycerin needed varies between 10-20%, depending on moisture content. The glycerin must be uniformly incorporated in food before freezing.

If part of the moisture on the surface of food is removed by drying, a smaller amount of glycerin is needed.

FORMULA NO. 1

Corn Kernels	100
Glycerin C.P.	15.2

No. 2

Chopped Meat	100
Glycerin C.P.	20

No. 3

Strawberries	100
Glycerin C.P.	21

No. 4

Fish, Potatoes or Noodle Mix	100
Glycerin C.P.	20

Chapter IX

POLISHES

Auto Polish

FORMULA NO. 1

(Spray Wax)

Mineral Seal Oil	20
"Tomah" Emulsifier Four	15
"Triton" X-100	1
Butyl "Cellosolve"	3
Water	61

NO. 2

(Cleaner)

"Viscasil" 10M	2.0
"SR-707" (40%)	4.8
"SR-131" (70%)	1.2
"Span" 80	2.0
"Tween" 81	0.3
Carnauba Wax Emulsion (20%)	5.0
Mineral Spirits	20.0
Kerosene	7.0
Water	47.5
"Kaopolite" SFO	10.0

No. 3

(Cleaner)

A	"Dow Corning" 200 Fluid (100 cs)	1.5
	"Dow Corning" 200 Fluid (1000 cs)	4.5
	Carnauba Wax #1 Yellow	1.0
	Beeswax	1.0
	Oleic Acid	2.5
	Kerosene	2.0
	Stoddard Solvent	18.0
B	Morpholine	1.5
	Water	16.0
C	"Snowfloss"	7.0
	"Celite" 315	7.0
	"Carbopol" 934	0.1
	Water	37.9

Heat A ingredients to 79 C (175 F) to melt waxes (Warning: Flammable). Mix B and C separately. While blending A with a high-shear mixer, add B slowly. Maintain temperature between 71 and 79 C (160 and 175 F). Remove heat, then add C while stirring.

No. 4

(Aerosol)

A	"Veegum"	1.5
	Sodium Carboxymethylcellulose (med. visc.)	0.2
	Water	48.0
B	Carnauba Wax	2.0
	Beeswax	2.0
	Oleic Acid	3.0
	Silicone (350 cs)	5.0
C	Morpholine	2.0
	Water	14.3
D	Mineral Spirits	20.0

POLISHES 319

E	Abrasive	2.0
	Preservative	q.s.

Dry blend the "Veegum" and CMC and add slowly to the water, agitating continually until smooth. Heat to 90 C. Melt B, stirring until uniform. Heat C to 60–70 C and add to B with mixing, holding temperature at 85-90 C. Heat D to about 70 C and add to B and C with mixing. Add A with rapid agitation. Continue mixing until cool. Add E to other components and mix until uniform.

Aerosol package: Concentrate 90%; Propellant 12/114, 60/40 10%.

No. 5

(Paste)

A	"Veegum"	1.0
	Water	29.0
B	Carnauba Wax	5.0
	Beeswax	3.0
	Ceresin	3.0
	"A-C" Polyethylene 629	2.5
	Silicone (350 cs)	5.0
	Stearic Acid	6.0
C	Morpholine	2.0
D	Naphthol Spirits	36.0
E	Abrasive	7.5
	Preservative	q.s.

Add the "Veegum" to the water slowly, agitating continually until smooth. Heat to 95 C. Melt B. Add C and maintain temperature at 90 C. Add D to B and C. Add A with vigorous agitation. When temperature reaches 50 C, add E and continue mixing until cool.

No. 6

(Detergent Resistant)

A	"Veegum" T	1.0
	Water	47.0
B	"Hoechst" Wax S	1.3
	"Hoechst" Wax E	0.2
	Silicone No. 530 Fluid	0.5
	Silicone No. 531 Fluid	3.0
	Sorbitan Monostearate	1.0
	Mineral Spirits	36.0
C	"Snow Floss"	10.0
	Preservative	q.s.

Add the "Veegum" to the water slowly, agitating continually until smooth. Heat to 80 C. Heat B to 75 C with stirring; avoid open flame. Add B to A with continuous agitation. Add C to A and B and continue mixing until cool.

Boat Liquid Polish/Restorer

"Viscasil" 10M	1.0
"SF-96" (1000)	6.0
"SR-131" (70%)	1.5
Oleic Acid	2.0
Mineral Spirits	15.0
Kerosene	13.0
Morpholine	1.0
"Carbopol" 934	0.1
Triethanolamine	0.1
Water	48.3
"Kaopolite" SF	8.0
"Super Floss"	4.0

Floor Polish

FORMULA No. 1

"Hoechst" Wax S	0.75
Carnauba Wax	1.75
"Tegiloxan®" (visc. 1000–5000 cP)	0.25

Silicone EL 49	6.00
White Spirit	51.25
Petrol (100–140)	40.00

No. 2

"Hoechst" Wax OM	1.50
Carnauba Wax	0.50
Beeswax	0.50
"Tegiloxan®" (visc. 1000–5000 cP)	0.25
Silicone EL 49	6.00
White Spirit	51.25
Petrol (100–140)	40.00

No. 3

"Syntran" 1026 (@ 15% N.V.)	85.0
"Syntran" 1501 (@ 15% N.V.)	15.0
Tributoxyethyl Phosphate	1.00
Methyl "Carbitol"	2.75
Ethylene Glycol	1.00
1% "FC"-120 or "FC"-128 Sol'n.	0.25

Adjust pH to 8.5 ± .2 with ammonia.

No. 4

(Non-Buffable)

Water	29.49
"FC"-128 (1%)	0.80
"SWS"-211	0.02
Diethylene Glycol Monomethyl Ether	2.00
Tributoxy Ethyl Phosphate	0.80
Formalin (37%)	0.15
"Rhoplex" B-1339 (38%)	53.94
"Zecolac" 802 (25%)	3.00
AC-392 (nonionic 35%)	3.57
AC-392 (anionic 25%)	10.00

No. 5

(Liquid Solvent)

"Petrolite" PE-100	3
"Petrolite" C-700	4
Paraffin Wax (145 F m.p.)	4
Stoddard Solvent	89

This type of suspension is prepared by dissolving the wax ingredients in one-half of the solvent at a temperature of 210–220 F. The remainder of the cold solvent is added to the wax solution with stirring. Stirring is continued until the dispersion or suspension reaches 90 F.

Floor Wax Stripper

Water	78.0
Tetrapotassium Pyrophosphate (anhydrous)	4.0
Sodium Metasilicate Pentahydrate	4.0
"Surco" A 10MM	4.0
Butyl "Cellosolve"	4.0
Monoethanolamine	4.0
"Surco" 233	2.0

Furniture Polish

FORMULA NO. 1

(Aerosol)

A	"Dow Corning" 200 Fluid (60,000 cs)	0.8
	"Dow Corning" 200 Fluid (350 cs)	3.2
	"Emcol" 14	0.8
	VM&P Naphtha	11.2
	Lemon Fragrance	0.4
B	"Co-Wax" Emulsion* (10%)	12.8
	Water	50.8
C	Propellant 12	20.0

Mix A until uniform. Add B with good agitation. Package.

*"Co-Wax" dilution—heat 10 parts wax to 90 C (194 F), heat 90 parts water to 90 C (194 F), mix with good agitation.

No. 2

(Emulsion)

A	"Dow Corning" 200 Fluid (500 cs)	3
	Linseed Oil	9
	"Igepal" CO-360	5
	VM&P Naphtha	35
B	Water	48

Add B to A with a high-shear mixer.

No. 3

(Cationic Emulsion)

A	"Dow Corning" 200 Fluid (350 cs)	3.00
	Carnauba Wax #1 Yellow	1.25
	"Be Square" Wax, White 190/195	0.25
	"Armac" 18D	2.75
	"Armac" HT	1.25
	Naphthol Mineral Spirits	26.00
B	Water	65.50

Heat A to 80 C (176 F) and B to 85 C (185 F). While blending A on a high-speed, high-shear mixer, slowly add B. Force-cool to 35 C (95F) in ice.

No 4

Concentrate:

Carnauba Wax	2.0
"Dow Corning" 200 Fluid (1000 cps)	3.0

"Isopar" E	20.0
"Witconol" F26-46	0.5
"Witconol" 14	0.5
Water	74.0

Aerosol:

Concentrate	80.0
Propellant 12	20.0

Mix the first five ingredients and heat to 80 C. Heat water to 80 C and add slowly with agitation. Cool to 55 C and pack.

Scratched Wood Polish

A	"Veegum" T	1.50
	Water	25.50
B	Stearic Acid	4.50
	Carnauba Wax	16.50
	Beeswax (yellow)	16.50
C	Triethanolamine	2.75
D	Mineral Spirits	27.00
E	Color	0.50
	Water	5.25

Add the "Veegum" T to the water slowly, agitating continually until smooth. Heat B to 90 C, add C and maintain temperature. Add D, maintaining 90 C, avoiding open flames. Add A to B, C, and D, mixing until uniform. Disperse E in hot water and add to the mixture, continuing to mix until uniform by processing through a roller mill.

Copper Cleaner

A	"Veegum" K	2.07
	"Kelzan"	0.23
	Water	78.55

B	"Snow Floss"	13.60
C	Buffer Solution*	q.s.
D	"Triton" X-102	4.65
	Benzotriazole	0.90
E	Perfume and Preservative	q.s.
	Color	q.s.

*Buffer solution: 1.46 parts–1 M H_3PO_4
1 part–125 g/l Na_3PO_4

Dry blend "Veegum" K and "Kelzan" and add to the water slowly, agitating continually until smooth. Add B to A gradually. Mix until smooth, then buffer this mixture to a pH of 2.5. Mix components in D until a clear solution is formed. Special care should be taken to avoid incorporation of air. Add D to other components very slowly and mix until uniform. Add E.

Glass Cleaner Polish

A	"Dow Corning" 200 Fluid (20 cs)	4.0
	Oleic Acid	2.5
	"Isopar" H	10.0
B	Morpholine	1.5
	Water	10.0
C	Water	22.0
D	"Snow Floss"	10.0
	Water	35.0
E	"Carbopol" 934 (2%)	5.0

While blending A on a high-shear mixer such as an Eppenbach, slowly add B. Add C, D, and E in order with continued agitation.

Black Shoe Polish

Concentrate:

Carnauba Wax	4.0
"Dow Corning" 200 Fluid (1000 cps)	3.0
"Witconol" F26-46	1.0
"Witconol" 14	0.7
Carbon Black (10% dispersion in "Isopar" E)	4.0
"Isopar" E	20.0
Water	67.3

Aerosol:

Concentrate	80.0
Propellant 12	20.0

Valve:

Vapor-tap valve with mechanical breakup actuator.

Prepare carbon black dispersion by mixing carbon black and "Isopar" E in a high-speed, high-shear mixer for 30 min.

Heat all formula ingredients together except water to 80 C. Heat water separately to 80 C and add to other ingredients with rapid agitation. Continue agitating while cooling. Pour into aerosol containers at 55-60 C.

This polish is sprayed onto shoe surfaces, spread with a cloth, then buffed with a brush.

Chapter X

TEXTILE SPECIALTIES

Bleaching Cotton

	g/l
Hydrogen Peroxide (35%)	20–30
Sodium Silicate	7.5
Caustic Soda	0.3
"Jaywet®" P-7	0.5
Optical Brightener	if desired

Start at room temperature and heat over 1 h to 190 F. Hold 1–2 h. cool and rinse.

Bleaching Polyester and Cotton

	g/l
Sodium Chlorite	3.0
Oxalic Acid	3.0
Odor and Corrosion Inhibitor	2.0
Optical Brightener	if desired

Sodium chlorite, when acidified gives off a toxic gas (chlorine dioxide). Care must be taken to use either closed equipment or adequate ventilation.

Start at room temperature and heat to 190–200 F over 30 min. Hold 1 h. Cool and give antichlor as follows:

Run 2 g/l sodium bisulfite for 20 min at 140 F followed by two warm rinses.

Note: It is always more efficient to bleach continuously by the rapid method if the equipment is available and the yardage to be processed large enough.

Textile Scouring Bath

"Jayscour®" K2X	2.0 g/l
Soda Ash	1.0 g/l

Run for 15–20 min at 190 F.

The soda ash does a much better job of removing lubricants and oils than does tetrasodium pyrophosphate owing to its higher alkalinity. Temperatures of 180–190 F should be employed since most waxes and lubricants normally used in knitting polyester/cotton blend fabrics have an emulsion and/or melting point of 160 F and temperatures below that level do not always insure complete removal of these products. If temperatures of less than 190 F can be employed satisfactorily, this is all the better.

Textile Wash-Wear Process

Acetic Acid	0.1
"Protowet" 100	0.1
"Protolube" L-20	3
"Protolube" HD	3
"Superez" AVG	20
"Curite" HC	4

"Superez" AVG is applied by the usual pad-dry-cure technique employing conventional equipment. "Superez" AVG is easily diluted with water, followed by the required auxiliary finishing agents. When the mix bath is almost to volume, the previously diluted catalyst is added. The mix bath should be made at approximately 100 F (38 C). Some heat is not detrimental to the stability of the bath.

The amount of "Superez" AVG will be determined by fabric type, blend level, etc. Generally, for a 50/50 polyester/cotton fabric—woven or know—approximately 20% "Superez" AVG, on weight of the bath, will give excellent performance.

Catalyst concentrations for 15-20% bath concentration of "Superez" AVG are as follows:

"Curite" PDQ (use with afterwash)	−3-4% on weight of bath
"Curite" 4403	−3-4% on weight of bath
"Curite" HC	−3-5% on weight of bath
"Curite" MG	−4-5% on weight of bath
"Curite" W	−4-5% on weight of bath
"Curite" 33	−3-5% on weight of bath

Curing time and temperature are approximately as follows:

300 F	3 min
340 F	1½ min
400 F	15-20 s

We recommend using the higher amount of catalyst when curing at the lower temperature. An afterwash is not necessary. If an afterwash is customary, a mild alkaline wash with peroxide, soda ash, and a detergent may be used.

Fabric Softener

"BTC®" 2125M P-40	50
"Ammonyx" 2200 P-100	20
Urea	30

Coning Oils

	% Emulsifier in Polybutene
	"Amoco" L-14 Polybutene
"Span" 40/"Tween" 40	20
"Span" 40/"Tween" 20	20
"Span" 40/"Tween" 60	20
"Span" 40/"Tween" 80	20
G-1186/"Tween" 20	20
G-1186/"Tween" 40	20
G-1186/"Tween" 60	20
G-1186/"Tween" 80	20
"Brij" 72/"Tween" 60	20

	"Amoco" H-300 Polybutene
"Brij" 78/"Span" 20	30
"Brij" 78/"Span" 60	30
"Span" 40/"Tween" 40	20

	"Amoco" H-1900 Polybutene
"Brij" 78/"Span" 20	50
"Span" 40/"Tween" 40	50

Stabilizing Cotton Corduroy

The stabilization of cotton corduroy by locking the pile is accomplished by application of a backcoating. The coating weight depends on fabric weight and on desired strength and fabric hand.

Water	66.2
"Nopco" DF-160L (diluted 1 : 1 with water)	0.2
"Rhoplex" TR-520	22.8
Ammonium Nitrate (25% sol'n.)	2.0
"Triton" GR-5	0.5
Ammonium Hydroxide (28% NH_3)	0.9
"Acrysol" ASE-60 ⎫ Premix	3.7
Water ⎭	3.7

The components should be added in the indicated order; an initial viscosity of about 2000 cps will be obtained. It can be applied by roller coating, then dried and cured. The resulting fabric will have a soft, full hand. Adjustments may be required in the ratio of "Rhoplex" TR-520 to "Acrysol" ASE-60, depending on mode of application, to control viscosity and penetration.

Print-Bonding Nonwoven Fabrics

(Improved Flame Resistance)

"Rhoplex" TR-520	44.5
Ammonium Hydroxide (28% NH_3)	0.3

Ammonium Sulfamate (60% sol'n.)	37.0
Ammonium Hydroxide (28% NH$_3$)	0.9
"Cellosize" QP 4400 (4% sol'n.)	17.3

Additions should be made in the indicated order. The ammonium sulfamate solution in particular should be added carefully in small increments. At this high level of salt, "Cellosize" QP-4400 has been found to be the preferred thickener. Final pH of the composition is 8.5 and the initial viscosity is 3310 cps. (Brookfield LVF Viscometer #2 spindle, 30 rpm.)

Flocking Compound

"Rhoplex" TR-520	85.9
"Flexol" TOF	3.4
"Nopco" DF-160L (diluted 1 : 1 with water)	0.15
"Cellosize" QP-4400	0.9
Water (cold)	8.6
Ammonium Hydroxide (28% NH$_3$)	1.1
pH	7.7
Viscosity (#4 spindle, 6 rpm), cps	70,000

This composition is applied by knife/roll coater to a cotton fabric and rayon flock is then applied. The fabric is cured at 300 F for about 4 min.

Textile Flame Retardants

FORMULA NO. 1

Dissolve sufficient "Antiblaze" 19 flame retardant in water to give a wet pickup of 3-6% flame retardant based on original fabric weight. Adjust the solution pH to 6.0 using disodium phosphate. Dissolve 0.02-0.05% nonionic or anionic surfactant if required.

Aqueous solutions prepared from "Antiblaze" 19 or "Antiblaze" 19T flame retardants retain their textile treating efficiency for over one week when stored at room temperature.

Remove all residual size and oils. Pad to 40-60% wet pickup. Dry for 1-2 min at 225-275 F. Thermosol 1-2 min at 365-400 F. Rinse thoroughly. Dry. Cure at 350-401 F.

No. 2

(For Tents)

Water	73.99
Wetting Agent	0.01
"Fyrol" 76	22.00
"Aerotex" M-3	3.00
Sodium Persulfate	1.00

No. 3

(For Polyester Fabric)

Water	66.0
"Fyrol" 76	25.0
"Aerotex" M-3	3.0
Wetting Agent	0.1
Sodium Alginate (2%)	0.3
Padding Emulsion	1.0
Pigments	2.0
Sodium Persulfate	2.0
Ammonium Hydroxide	0.6

No. 4

(For Army Duck)

Water	60.63
"Fyrol" 76	30.00
"Aerotex" M-3	4.00
Wetting Agent	0.25
Sodium Alginate (2%)	1.25
Latex Binder	1.00
Pigments	0.87
Sodium Persulfate	2.00

Mildewicide, Waterproof

Water	88.5
"Pentel" GH-28	6.0
"Aerotex" WR-96	4.0
Catalyst X-4	1.0
"Merkyl" PM-TL	0.5

Cleaning Dyeing Equipment

Fill the machine 2/3 full with water and add:

"Merse" RTD	3 g/l
"Tanapon" Clear	6 g/l

or

"Merse" RTD	3 g/l
Caustic Soda	3 g/l
Sodium Hydrosulfite	3 g/l

Fill to capacity and heat to 250-265 F. Hold for 30 min. Blow down if the equipment installation will allow. If not, cool back to depressurizing temperature (normally 190-200 F) and drop bath. Give running wash with warm water to flush out residues.

For package machine cleaning, set the empty yarn carriers into the package machine. This simultaneously cleans the inside of the carrier arms.

Cotton Mill Dust Control

Spray 1% "Milube" N-32 on loose cotton.

Chapter XI

MISCELLANEOUS

Foliar Trace Element Spray

"Rayplex"	1 lb
Water	10–12 gal

Agricultural Pesticides

FORMULA NO. 1

(Flowable Sulfur)

A	"Veegum" T	0.65
	Water	35.17
B	"Triton" X-100	0.08
	"Darvan®" No. 1	1.26
	Glycerin	4.14
	Sulfur	52.50
	Urea	6.20

Add the "Veegum" T to water by mixing with high shear equipment (such as a homomixer) until gel is formed and is smooth. Transfer to suitable equipment (such as homomixer or ball mill) and add the ingredients in B in the order shown. Ball mill or stir at high speed in homomixer until uniform.

Urea and glycerin are included for freeze-thaw stability.

No. 2

(Flowable Copper Sulfate-Sulfur)

A	"Veegum" T	0.20
	Water	35.85
B	"Cellosize" 52000 H	0.25
	Methyl Cellulose (15 cps)	0.05
	"Triton" X-100	0.15
	"Darvan®" No. 1	1.00
	"Darvan®" No. 4	1.50
	Tribasic Copper Sulfate	16.20
	Sulfur	32.30
	Urea	7.50
	Glycerin	5.00

Add the "Veegum" T to water by mixing with high shear equipment (such as homomixer) until smooth. Transfer to suitable equipment (such as a homomixer or ball mill) and add the ingredients in B. Ball mill or stir at high speed in homomixer until uniform.

		No. 3	No. 4
		(Flowable Atrazine)	
A	"Veegum" T	0.6	0.5
	Water	49.0	53.1
B	"Triton" X-100	0.1	0.1
	"Darvan®" No. 1	1.3	1.3
	Ethylene Glycol	4.0	—
	Atrazine	45.0	45.0

Add the "Veegum" T to the water by mixing with high shear equipment (such as homomixer) until a smooth gel is formed. Transfer to suitable equipment (such as homomixer or ball mill) and add the ingredients in B. Ball mill or stir at high speed in a homomixer until uniform.

Antifoam agents or vacuum mixing equipment may be desired if excessive foaming occurs.

		NO. 5	NO. 6
		(Flowable Sevin)	
A	"Veegum" T	0.7	0.90
	Water	31.7	70.20
B	"Triton" X-100	0.1	0.05
	"Darvan®" No. 1	1.3	0.65
	Glycerin	4.1	2.00
	Sevin	56.0	23.00
	Urea	6.2	3.20
	Citric Acid	0.1	0.10

Add the "Veegum" T to the water by mixing with high shear equipment (such as homomixer) until gel is formed and is smooth. Transfer to suitable equipment (such as homomixer or ball mill) and add the ingredients in B in the order shown. Ball mill or stir at high speed in homomixer until uniform.

Varied Pesticide Suspensions

FORMULA	NO. 1	NO. 2	NO. 3	NO. 4	NO. 5
Active Pesticide	48.0	48.0	48.0	48.0	48.0
Propylene Glycol	3.0	3.0	3.0	3.0	3.0
"Pluraflo" E4 or E5	2.0	3.0	4.0	5.0	6.0
Gelling Clay	4.0	2.0	1.0	0.5	—
Water	43.0	44.0	44.0	43.5	43.0

After they have been prepared and allowed to stand for sufficient time to develop the thixotropic properties, the formulations can be evaluated to determine which merit further development.

Personal Insect Repellent

FORMULA NO. 1

N,N-Diethyl-*m*-Toluamide	13.15
Other Isomers	.70
"MGK" 264	3.96

"MGK" 11	.99
"MGK" 326	.99
Deodorized Kerosene	80.21

NO. 2

"Poloxamer" 407 or Poloxamine 1508	14
Isopropyl Alcohol	12
N,N-Diethyl-*m*-Toluamide	10
Water	64
Preservative	q.s.

NO. 3

"Tetronic" 1508	14
Isopropyl Alcohol	12
N,N-Diethyl-*m*-Toluamide	10
Water	64
Preservative	q.s.

Place surfactant, alcohol, water and amide into suitable container. Warm to 80 C, mix gently until the surfactant has dissolved and the mixture is homogeneous. Add preservative and transfer to suitable containers. Product sets up into a clear, ringing gel when cooled to room temperature.

NO. 4

"Polawax"	3.0
N,N-Diethyl-*m*-Toluamide	15.0
Ethanol SDA-40 (95%)	39.6
Water	32.4
Propellant 12/114 (40 : 60)	10.0

Fly Bait Poison

Sugar	12
Malathion (25% emulsion)	2–4
Water	to make 128

Roach Killer

Orthoboric Acid	95
"Aerosil"	5

Cricket Control

Coarse Bran	100
Molasses (crude)	15
Sodium Fluosilicate	5
Water	120

Termite Control

Aldrin in an Oil or Water Emulsion	0.5
Gammexane in an Oil or Water Emulsion	0.8
Chlordane in an Oil or Water Emulsion	0.5
Dieldrin in an Oil or Water Emulsion	0.5
Lindane in an Oil or Water Emulsion	0.8
DDT in an Oil-Based Emulsion	8.0

Trichlorbenzene in oil at a ratio of 1.3 parts oil

Pentachlorophenol Emulsion	5

These can be applied at the rate of 2 gal per five linear feet.

Moths and Other Wool Digesting Insects

FORMULA	NO. 1	NO. 2	NO. 3	NO. 4	NO. 5
Pyrethrins	0.04	0.025	0.25	—	0.025
Piperonyl Butoxide	0.20	0.20	0.2	—	0.2
"Thanite"	—	1.0	—	2.5	—
"Lethane"	—	—	1.5	—	—
Dieldrin	0.50	0.06	0.06	0.06	0.06
Perfume and Odorless Kerosene up to	100	100	100	100	100

Rat Biscuit

FORMULA NO. 1

Corn Meal (White)	1 lb
Peanut Butter	1 oz
Molasses	2 oz
Barium Sulfate	2 lb

Mix well; cut into discs; bake at 350 F for about 20 min.

NO. 2

(Nonpoisonous)

Portland Cement	1 lb
Vegetable Oil	2 oz

Mix the above well. Add

Peanut Butter	2 oz
Starch	1 oz

Mix well; form into cookies; bake at 350 F for 20 min.

Grain Fumigant

Carbon Tetrachloride	80
Carbon Disulfide	20

Heat Transfer Printing Ink

FORMULA	NO. 1	NO. 2	NO. 3	NO. 4
Cellulose Acetate Propionate	10.0	10.0	10.0	10.0
"Tecsol®" C	59.5	76.5	–	–
Ethyl Acetate (99%)	25.5	8.5	–	–
Isopropyl Alcohol	–	–	68.0	–
n-Propyl Alcohol	–	–	–	68.0
Water	–	–	17.0	17.0
"Eastman®" HTP Red FFBL Dye	5.0	5.0	5.0	5.0

One-Time Carbon Ink

"Petrolite" WB-5	15.0
Paraffin Wax (156 F, 68.8 C)	30.0
Methyl Violet Base	0.1
Furnace Black	22.0
Toning Iron Blue	3.0
Mineral Oil (100 SUS)	29.9

Charge waxes and dyes in a steel ball mill at 200-215 F (93.3-101.6 C). Mill until dyes are in solution (15-30 min). The balance of the materials are added to the ball mill and milled for 2½ h.

Preserving Newspapers

Milk of Magnesia	4 oz
Carbonated Water	1 qt

Soak the flattened newspaper clipping in the above. Drip dry and pat dry.

Investment Casting Wax

"Petrolite" WB-10	35
Paraffin Wax (m.p. 140-142 F–60-61.1 C)	35
Alpha Methyl Styrene Resin	30

Preventing Wax Candle Discoloration

White candles will requently discolor on short light exposure, depending on the degree of refinement of the paraffin and stearic acid used. This discoloration may usually be controlled by as little as 0.05% "Tinuvin" P. Light stability of candle colors is also improved, as is the thermal stability of several wax-soluble dyes.

Cationic Wax Emulsion

"Ceramer" 67	60
"Igepal" CO-610	2
"Ethoquad" 18/12	18
Water	320

Petroleum Wax Emulsion

"Petrolite" C-9500	40
"Petrolite" C-700	40
Emulsifiable Polyethylene	20
Oleic Acid	2
"Triton" N-101	1
KOH (85%)	2
Morpholine	6
Water to desired solids	

Vegetable Oil Emulsion

(W/O)

Vegetable Oil	80.00
"Generol" 122	5.00
Water	15.00

Solubilized Essential Oils

FORMULA NO. 1

(Lavender)

Polyglycol (30) Glyceryl Laurate	7
Lavender Oil	1

NO. 2

(Peppermint)

Polyglycol (30) Glyceryl Laurate	7
Peppermint Oil	1

Encapsulating Essential Oils

Refined Gum Arabic	1.3
Water	74.3
Essential Oil	14.1
Malto-Dextrin	10.3

Using high-speed agitation, disperse the gum arabic, slowly, into the vortex of the preheated water (140 F) and allow to dissolve (approximately 20 min). Add the malto-dextrins to the solution with agitation. Add the essential oil with agitation to form crude dispersion.

Homogenize on a two-stage homogenizer:

First stage	2000 psig
Second stage	1000 psig

Emulsion can be spray dried at inlet temperatures of 350–400 F and outlet temperatures of 200–220 F.

Lubricating Emulsion

"Petrolite" WB-10	100
"American Hoechst" Emulsifier 2106	20
Diethylethanolamine	6
Water (to 14% solids)	606

Melt the wax and the emulsifier in the same container and raise the temperature to 235–240 F (112.7–115.5 C). Add the DEAE and react for 5 min with agitation. Maintain the above temperature. Heat the water to 205F (96.1 C) and add the molten wax and emulsifier system to the water with good agitation. Cool the emulsion to room temperature.

Clear Mineral Oil Emulsion

	Formula No. 1	No. 2
"Crodafos" N10 Neutral	4.0	—
"Crodafos" N3 Neutral	—	5.0
"Volpo" 20	—	1.0
"Volpo" 10	12.0	15.0
"Volpo" 3	4.0	—
"Carbowax" 200	10.0	—
Glycerin	—	10.0

Mineral Oil	20.0	12.0
Perfume, Preservative, and Color	q.s.	q.s.
Water	50.0	57.0

Melt the "Crodafos," "Polychol" and/or "Volpo" and other oil phase ingredients adding the glycerin or "Carbowax" to this mixture. Bring to 85-90 C and bring the water to the same temperature—provide for water loss during manufacture either by use of a covered mixing vessel or a slight excess of water, since it is difficult to qs the water after the gel has formed.

At that temperature, slowly add the water to the oils with moderate agitation to avoid air bubbles. When the addition is complete, begin cooling and continue agitation.

Mineral Oil Emulsion

FORMULA	NO. 1	NO. 2	NO. 3
80 s–SUS	66.7	66.5	–
50-60 s–SUS	–	–	65
Emulsifier			
"Lipal" TE-43	5.5	22.3	5.8
"Lipal" CE-55	–	11.2	–
"Lipal" CE-71	16.7	–	17.5
"Drewmulse" GMO	11.1	–	11.7

NO. 4

"Manucol" DH	0.45
Mineral Oil	50.0
Gum Tragacanth	0.5
Color and Flavor	as required
Preservative	as required
Water	to 100

Disperse the "Manucol" DH in the oil. Dissolve the other ingredients in the water. Add the "Manucol" DH slurry to the water and mix thoroughly. Homogenize the mixture.

Lower Alcohol Fatty Ester Emulsion

FORMULA	No. 1	No. 2	No. 3
Isopropyl Myristate	62	66.0	—
Butyl Stearate	—	—	70
Emulsifiers			
"Lipal" TE-43	20	—	20
"Lipal" TE-70	18	16.0	10
"Lipal" CE-55	—	9.0	—
"Drewpol" 10-4-0	—	9.0	—

Triglyceride Emulsion

FORMULA	No. 1	No. 2	No. 3
Coconut Oil	70	67.7	—
Triolein	—	—	70
Emulsifier			
"Lipal" TE-43	10	—	—
"Lipal" TE-70	20	—	20
"Lipal" CE-55	—	21.5	—
"Lipal" 300-W	—	—	10
"Drewmulse" GMO	—	10.8	—

All the above form stable 10–20% oil in water emulsions.

Defoaming Agents

FORMULA No. 1

Pine Oil/Dipentene	95.0
Aluminum Stearate or Zinc Stearate	5.0

No. 2

Pine Oil/Dipentene	97.0
Calcium Oleate	3.0

No. 3

Pine Oil	78.75
Higher Alcohols	26.25
Aluminum Stearate	5.00

Fuel Oil Antistatic

Add 0.5–5 lb "Tolad" 311 to fuel oil.

Fuel Oil Improver

50–500 ppm "Tolad" 35 is added to fuel oil.

Fuel Oil Sludge Dispersant

Add 1 gal "Tolad" 316 to 4000 gal fuel oil.

Antifreeze

Borax	7
Disodium Hydrogen Phosphate	16
Mercaptobenzothiazole	14
Water	80
Ethyleneglycol	80

Mix 1–2 oz of above to	
Ethyleneglycol	1 gal

Preventing Icing on Refrigerator Coils

Apply coating of petrolatum to all surfaces. Moisture that condenses will run off or, when it freezes, it can be scraped off easily.

Decreasing Sensitivity of Ammonium Nitrate

Add 2 lb "Armoflo" 65 per ton of ammonium nitrate.

Highway Pavement

Sulfur	2.5-3
Asphalt	2.5-3
Aggregate	to make 100

This shows improved low temperature and fatigue resistance.

Glass Bottle Protector

"Myrj" 52-5	0.02-1
Water (hot)	99

Spray on bottle and allow to dry. This coat acts as a bumper and prevents breakage.

Wood Preservative Treatment

1. Preservative treatment.

Aluminum Stearate	2	
Rosin Ester	8	Apply as 5 min dip or
Pentachlorophenol	5	by pressure treatment.
"Solvesso" 100	to 100	Dry thoroughly.

2. Apply 1 coat wash primer. Air dry 15-30 min.

3. Apply 2 coats high grade alkyd paint (med. or long oil). Air dry or bake. Note: can use new bake system of 30 min at 190 F, very fast, will not harm wood. This is generally for pigmented systems.

Removal of Toxic Gases from Smoking Tobacco

FORMULA NO. 1

Tobacco	100
Potassium Chlorate (10% water sol'n.)	10
Copper Hydroxide	1-2
Diethylene Glycol	5

NO. 2

Tobacco	100
Potassium Chlorate (10% water sol'n.)	10
Glycerin	3
Manganese Hydroxide	2-4

NO. 3

Tobacco	100
Potassium Chlorate (10% water sol'n.)	10
Triethyleneglycol	4
Cobalt Hydroxide	2-4

NO. 4

Tobacco	100
Potassium Chlorate (10% water sol'n.)	10
Sorbitol	5
Water	4
Ferrous Hydroxide	2-4

NO. 5

Tobacco	75
Active Carbon	25
Potassium Chlorate (10% water sol'n.)	10
Diethyleneglycol	4
Cuprous Hydroxide	2-4

Tobaccoless Snuff

FORMULA NO. 1

Powdered Roasted Rice Hulls	34
Powdered Cellulose	22
Water	37
Caffeine	2

NO. 2

Powdered Cellulose	92.9
Synthetic Tobacco Flavor	2
Glycerin	3
Sorbic Acid	0.1
Caffeine	2
FDC Brown Color	q.s.

NO. 3

Powdered Peanuts	34
Powdered Cellulose	22
Water	37
Powdered Caffeine	2
Tobacco Extract	2
Glycerin	3
Potassium Sorbate	0.1
FDC Brown Color	q.s.

NO. 4

Powdered Alumina	34
Powdered Cellulose	22
Water	37
Caffeine	2
Synthetic Tobacco Flavor	2
Glycerin	3
Sorbic Acid	0.1
FDC Brown Color	q.s.
Citric Acid	0.9

No. 5

Wheat Bran	34
Powdered Cellulose	32
Tobacco Extract (oleoresin)	32
Glycerin	3
Sorbic Acid	0.1
Tartaric Acid	0.9
Rose Geranium Oil	0.1
Water	37
Caffeine	2

Copper Antitarnish

A	"Veegum" K	2.07
	"Kelzan"	0.23
	Water	78.55
B	"Snow Floss"	13.60
C	Buffer Solution*	q.s.
D	"Triton" X-102	4.65
	Benzotriazole	0.90
E	Perfume and Preservative	q.s.
	Color	q.s.

*Buffer solution: 1.46 parts–1 M H_3PO_4
1 part–125 g/l Na_3PO_4

Dry blend "Veegum" K and Kelzan and add to the water slowly, agitating continually until smooth. Add B to A gradually. Mix until smooth, then buffer this mixture to a pH of 2.5. Mix components in D until a clear solution is formed. Special care should be taken to avoid incorporation of air. Add D to other components very slowly and mix until uniform. Add E.

Tarnish Remover

Thiourea	50 lb
Sulfuric Acid	1 gal
"Triton" X-100	1 gal
Water	96 gal

Aluminum Etch

H_2SO_4	150 ± 50 g/l
$Na_2Cr_2O_7$	40 ± 20 g/l
Temperature	60 ± 5 C
Immersion Time	30 min

Removing Metal Substrates

Metal Substrate	*Common Acid Solutions**
Aluminum	25-50% nitric acid, by volume
Copper, Copper Alloys	5-15% H_2SO_4 or HCL, by volume
High-Carbon Steel	Anodic treatment in 20-65% H_2SO_4, 15-40 A/dm^2, 1 min
Leaded Bronze	5-25% HBF_4
Stainless Steels	1-2% H_2SO_4 and 0.1% HF, 1-2 min
Steel	5-10% H_2SO_4, by volume, 15 s-2 min**
Zinc Die Castings	0.25-0.75% H_2SO_4, 20-30 s

*All solutions at room temp
**A 3-6% H_2SO_4 sol'n. heated to 70-80 C is commonly used for removing heavy scale.

Industrial Water Rust Inhibitor

Zinc Gluconate	0.1-0.5%

in cooling water.

Metal Forming Lubricant

Tallow Soap	37
Mineral Oil	27
Water	36

Add oil to water with good mixing.

Mold Release Agent

"Emphos" D70-30C	5.0
Propellants 12/11 (10 : 90)	95.0

Antistatic Spray

"Emcol" CC-36	25
Isopropanol	75
Propellant	400

Air Fresheners

FORMULA	NO. 1	NO. 2	NO. 3	NO. 4	NO. 5
"Meelium" (100% active basis)	1.2	0.1	1.5	1.6	1.6
Compatible Perfume Oil	0.5	13.7	0.5	0.2	0.3
Deodorized Kerosene*	13.3	–	–	18.2	18.1
1,1,1-Trichloroethane	–	42.5	49.0	–	–
Propellant 11	42.5	42.5	–	40.0	40.0
Propellant 12	42.5	1.2	49.0	40.0	40.0

*Or other suitable diluent.

Sodium Bisulfite Gel

"Pluronic" F-127 Polyol	22.0
Sodium Bisulfite	2.2
Water	75.8

Dissolve "Pluronic" F-127 polyol in cold water (5-10 C). Then add sodium bisulfite and mix until dissolved. Transfer to containers. A clear, ringing gel with a pH of 5.5 forms on warming up to room temperature.

APPENDIX

Federal Laws Regulating Foods, Drugs, and Cosmetics

Anyone who plans to market a food, drug, cosmetic, or chemical specialty should be thoroughly familiar with the Federal laws and regulations pertaining to that particular product.

The Federal Food and Drug Law, which was passed in 1938, was comparatively lenient in its restrictions. But the discovery of more potent drugs, new food additives, and powerful insecticides posed safety problems. In some instances, dangerous toxicity of a new compound was not suspected prior to marketing; then, when its use was widespread, serious poisonings and other deleterious effects were reported.

In 1962, amendments to the 1938 law were passed. These amendments were very restrictive, demanding far more rigid proof of the safety of any drug or chemical used on or by humans; they also brought under Federal control products used on or consumed by both pet and food animals.

Responsibility for enforcement of these laws is still divided among three government agencies. The Food and Drug Administration (FDA), a bureau of the Department of Health, Education, and Welfare, has the broadest regulatory powers. This agency controls the marketing of all food, medications, and cosmetic products for human use as ministration (FDA), a bureau of the Department of Health, Education, and Welfare, has the broadest regulatory powers. This agency controls the marketing of all foods, medications, and cosmetic products for human use as well as of all veterinary medications.

The Federal Trade Commission (FTC) is charged with the responsibility of policing advertising, to check false, misleading, or illegal statements. One exception is drug advertising to physicians, which comes under FDS jurisdiction.

The Department of Agriculture has jurisdiction over pesticides and their labeling and over additives to foods for animals being raised for human consumption.

Cosmetics

Cosmetics are not restricted as stringently as drugs, but the definition of a cosmetic is very narrow. Any preparation used *only* for cleansing or beautification of the skin, hair, or fingernails is considered a cosmetic. There are certain prohibited drugs or chemicals that are considered too dangerous for use in cosmetics, but on the whole it is up to the manufacturer to market products that are safe. Cosmetic claims can be quite flamboyant, so long as they are only for beautification. Any claim of a medicinal nature, however, even if it is only implied, immediately places the product in the drug class, which means that official permission to market the product is mandatory.

The law does not forbid the use of poisonous substances in a cosmetic. The ban applies only when the quantity of a poisonous ingredient *may* be injurious to users, *under the directions for use as given by the manufacturer.* The product may be extremely hazardous if misused, yet still be acceptable, under the law, *provided* safe directions are included and proper caution statements are made on the label.

While the FTC has charge of all cosmetic *advertising*, it is the FDA that controls *labeling*. The FDA has the authority to seize any product in interstate commerce that it considers misbranded. The term "labeling" is a broad designation; it covers much more than just the label pasted on the container. Any written, printed, or graphic material accompanying the package is considered part of labeling, even if it is delivered by mail, separately from the actual product. A display in a store, along with the product, is also "labeling": so is any literature handed out by demonstrators.

The manufacturer of a cosmetic is thus literally "walking a verbal tightrope," as the following *partial* list of taboos will further illustrate. Inclusion of any of these designations is sufficient to bring a charge of misbranding a cosmetic:

Circulating cream	Hair grower
Contour cream	Hair restorer or revitalizing
Deep pore cleanser	preparation
Depilatory for *permanent* removal of hair	Muscle oil
	Nail grower

Enlarged pore paste
Eyelash grower
Eye wrinkle cream
Scalp food
Skin conditioner or tonic
Skin food or nourishing cream
Skin texture preparation
Stimulating cream
Hair color restorer

Nonallergic product
Peroxide cream
Tissue cream
Wrinkle and double chin
  eradicator
Any representation that a
  product is superior in cosmetic value because it
  contains a vitamin

The law does not require that a cosmetic contain a list of its ingredients, but if the labeling puts it in the "drug" class, it must list all ingredients and state the quantities. Any statement that a cosmetic will cure or alleviate dandruff, promote a "healthy skin or scalp," alleviate itching, revitalize dry skin, or cure acne, eczema, freckles, rash, or any "skin trouble" places the product in the "drug" class.

The manufacturer or distributor of a cosmetic has the full responsibility of a truthfully labeled cosmetic, as defined by the FDA. The FDA is not authorized to approve the label or labeling of a cosmetic, but usually will give an informal opinion as to whether a proposed label is to be construed in fact as a drug label. Even such an informal opinion, however, does not relieve the manufacturer of his responsibility, because such an opinion is not an official acceptance of either the product or its labeling.

Additions and amendments to these laws are often made. It is best to write to the proper authority to get up-to-date information.

Some Incompatible Chemicals

The substances in the lefthand column must be stored and handled so that they cannot come into any contact with the substances in the right-hand column.*

Alkaline and alkaline-earth metals, such as sodium, potassium, cesium, lithium, magnesium, calcium, aluminum

Carbon dioxide, carbon tetra-chloride, and other chlorinated hydrocarbons. (Also prohibit water, foam, and dry chemical on fires involving these metals.)

---

*Based partly on *Dangerous Chemicals Code*, 1951 Edition, p. 19–20, Bureau of Fire Prevention, City of Los Angeles Fire Department, published by Parker & Company, Los Angeles 13, California.

Some Incompatible Chemicals (contd.)

Acetic acid	Chromic acid, nitric acid, hydroxyl-containing compounds, ethylene glycol, perchloric acid, peroxides, and permanganates.
Acetone	Concentrate nitric and sulfuric acid mixtures.
Acetylene	Chlorine, bromine, copper, silver, fluorine, and mercury.
Ammonia (anhydr)	Mercury, chlorine, calcium hypochlorite, iodine, bromine, and hydrogen fluoride.
Ammonium nitrate	Acids, metal powders, flammable liquids, chlorates, nitrites, sulfur, finely divided organics or combustibles.
Aniline	Nitric acid, hydrogen peroxide.
Bromine	Ammonia, acetylene, butadiene, butane and other petroleum gases, sodium carbide, turpentine, benzene, and finely divided metals.
Calcium carbide	Water (See also acetylene.)
Calcium oxide	Water.
Carbon, activated	Calcium hypochlorite.
Copper	Acetylene, hydrogen peroxide.
Chlorates	Ammonium salts, acids, metal powders, sulfur, finely divided organics or combustibles.
Chromic acid	Acetic acid, naphthalene, camphor, glycerol, turpentine, alcohol, and other flammable liquids.
Chlorine	Ammonia, acetylene, butadiene, butane and other petroleum gases, hydrogen, sodium carbide, turpentine, benzene, and finely divided metals.

Some Incompatible Chemicals (contd.)

Chlorine dioxide	Ammonia, methane, phosphine, and hydrogen sulfide.
Fluorine	Isolate from everything.
Hydrocyanic acid	Nitric acid, alkalis.
Hydrogen peroxide	Copper, chromium, iron, most metals or their salts, any flammable liquid, combustible aniline, nitro-methane.
Hydrofluoric acid, anhydrous (hydrogen fluoride)	Aqueous or anhydrous ammonia.
Hydrogen sulfide	Fuming nitric acid, oxidizing gases.
Hydrocarbons (benzene, butane, propane, gasoline, turpentine, etc.)	Fluorine, chlorine, bromine, chromic acid, sodium peroxide.
Iodine	Acetylene, anhydrous or aqueous ammonia.
Mercury	Acetylene, fulminic acid, ammonia.
Nitric acid (conc)	Acetic acid, aniline, chromic acid, hydrocyanic acid, hydrogen sulfide, flammable liquids, flammable gases, and nitritable substances.
Nitroparaffins	Inorganic bases.
Oxygen	Oils, grease, hydrogen, flammable liquids, solids or gases.
Oxalic acid	Silver, mercury.
Perchloric acid	Acetic anhydride, bismuth and its alloys, alcohol, paper, wood, grease, oils.
Peroxides, organic	Organic or mineral acids; avoid friction.
Phosphorus (white)	Air, oxygen.
Potassium chlorate	Acids (See also chlorate.)
Potassium perchlorates	Acids (See also perchloric acid.)
Potassium permanganate	Glycerol, ethylene glycol, benzaldehyde, sulfuric acid.

Some Incompatible Chemicals (contd.)

Silver	Acetylene, oxalic acid, tartaric acid, fulminic acid, ammonium compounds.
Sodium	See alkaline metals.
Sodium nitrate	Ammonium nitrate and other ammonium salts.
Sodium oxide	Water.
Sodium peroxide	Any oxidizable substance, such as ethanol, methanol, glacial acetic acid, acetic anhydride, benzaldehyde, carbon disulfide, glycerol, ethylene glycol, ethyl acetate, methyl acetate, and furfural.
Sulfuric acid	Chlorates, perchlorates, permanganates.
Zirconium	Prohibit water, carbon tetrachloride, foam, and dry chemical on zirconium fires.

Tables

Weights and Measures
Troy Weight

24 gr = 1 pwt
20 pwt = 1 oz
12 oz = 1 lb

Apothecaries' Weight

20 gr = 1 scruple
3 scruples = 1 dr
8 dr = 1 oz
12 oz = 1 lb

The ounce and pound are the same as in Troy Weight.

Avoirdupois Weight

27-11/32 gr = 1 dr
16 dr = 1 oz
16 oz = 1 lb
2000 lb = 1 short ton
2240 lb = 1 long ton

Dry Measure

2 pt = 1 qt
8 qt = 1 pk
4 pk = 1 bu
36 bu = 1 chaldron

Tables (contd.)

Liquid Measure

4 gi = 1 pt
2 pt = 1 qt
4 qt = 1 gal
31½ gal = 1 bbl
2 barrels = 1 hogshead
1 tsp = 1/6 oz
1 tblsp = 1/2 oz
16 fl oz = 1 pt

Square Measure

144 in.^{2} = 1 ft^2
9 ft^2 = 1 yd^2
30¼ yd^2 = 1 rod^2
43,560 ft^2 = 1 acre
40 rod^2 = 1 rood
4 roods = 1 acre
640 acres = 1 mile^2

Metric Equivalents
Length

1 in. = 2.54 c
1 ft = 0.305 m
1 yd = 0.914 m
1 mile = 1.609 km
1 cm = 0.394 in.
1 m = 3.281 ft
1 m = 1.094 yd
1 km = 0.621 mile

Capacity

1 U.S. fl oz. = 29.573 ml
1 U.S. Liquid qt = 0.946 l
1 U.S. dry qt = 1.101 l
1 U.S. gal = 3.785 l
1 U.S. bu = 0.3524 hl
1 in^3 = 16.4 m^3

Capacity (contd.)

1 m = 0.034 U.S. fl oz
1 ℓ = 1.057 U.S. liquid qt
1 ℓ = 0.908 U.S. dry qt
1 ℓ = 0.264 U.S. gallon
1 hl = 2.838 U.S. bu.
1 cm^3 = .061 in.^{3}
1 ℓ = 1000 ml or 1000 cm^3

Circular Measure

60 s = 1 min
60 min = 1 degree
360 degrees = 1 circle

Long Measure

12 in. = 1 ft
3 ft = 1 yd
5½ yd = 1 rod
5280 ft = 1 stat. mile
320 rods = 1 stat. mile

Weight

1 gr = 0.065 g
1 apoth. scruple = 1.296 g
1 av oz = 28.350 g
1 troy oz = 31.103 g
1 av. lb. = 0.454 km
1 troy lb. = 0.373 km
1 g = 15.432 gr
1 g = 0.772 apoth. scruple
1 g = 0.035 av oz
1 g = 0.032 troy oz
1 km = 2.205 av lb
1 km = 2.679 troy lb

ATOMIC WEIGHTS

(Alphabetical Order)

Element	Symbol	Atomic Number	Atomic Weight*
Actinium	Ac	89	(227)
Aluminum	Al	13	26.9815
Americium	Am	95	(243)
Antimony	Sb	51	121.75
Argon	Ar	18	39.948
Arsenic	As	33	74.9216
Astatine	At	85	(210)
Barium	Ba	56	137.34
Berkelium	Bk	97	(247)
Beryllium	Be	4	9.0122
Bismuth	Bi	83	208.980
Boron	B	5	10.811
Bromine	Br	35	79.909
Cadmium	Cd	48	112.40
Calcium	Ca	20	40.08
Californium	Cf	98	(251)
Carbon	C	6	12.01115
Cerium	Ce	58	140.12
Cesium	Cs	55	132.905
Chlorine	Cl	17	35.453
Chromium	Cr	24	51.996
Cobalt	Co	27	58.9332
Copper	Cu	29	63.54
Curium	Cm	96	(247)
Dysprosium	Dy	66	162.50
Einsteinium	Es	99	(254)
Element 102		102	(254)
Erbium	Er	68	167.26
Europium	Eu	63	151.96
Fermium	Fm	100	(253)
Fluorine	F	9	18.9984
Francium	Fr	87	(223)

*A value given in parentheses denotes the mass number of the longest-lived isotope.

Element	Symbol	Atomic Number	Atomic Weight*
Gadolinium	Gd	64	157.25
Gallium	Ga	31	69.72
Germanium	Ge	32	72.59
Gold	Au	79	196.967
Hafnium	Hf	72	178.49
Helium	He	2	4.0026
Holmium	Ho	67	164.930
Hydrogen	H	1	1.00797
Indium	In	49	114.82
Iodine	I	53	126.9044
Iridium	Ir	77	192.2
Iron	Fe	26	55.847
Krypton	Kr	36	83.80
Lanthanum	La	57	138.91
Lawrencium	Lw	103	(256)
Lead	Pb	82	207.19
Lithium	Li	3	6.939
Lutetium	Lu	71	174.97
Magnesium	Mg	12	24.312
Manganese	Mn	25	54.9380
Mendelevium	Md	101	(256)
Mercury	Hg	80	200.59
Molybdenum	Mo	42	95.94
Neodymium	Nd	60	144.24
Neon	Ne	10	20.183
Neptunium	Np	93	(237)
Nickel	Ni	28	58.71
Niobium	Nb	41	92.906
Nitrogen	N	7	14.0067
Osmium	Os	76	190.2
Oxygen	O	8	15.9994
Palladium	Pd	46	106.4
Phosphorus	P	15	30.9738
Platinum	Pt	78	195.09
Plutonium	Pu	94	(244)
Polonium	Po	84	(209)

*A value given in parentheses denotes the mass number of the longest-lived isotope.

Element	Symbol	Atomic Number	Atomic Weight*
Potassium	K	19	39.102
Praseodymium	Pr	59	140.907
Promethium	Pm	61	(145)
Protactinium	Pa	91	(231)
Radium	Ra	88	(226)
Radon	Rn	86	(222)
Rhenium	Re	75	186.2
Rhodium	Rh	45	102.905
Rubidium	Rb	37	85.47
Ruthenium	Ru	44	101.07
Samarium	Sm	62	150.35
Scandium	Sc	21	44.956
Selenium	Se	34	78.96
Silicon	Si	14	28.086
Silver	Ag	47	107.870
Sodium	Na	11	22.9898
Strontium	Sr	38	87.62
Sulfur	S	16	32.064
Tantalum	Ta	73	180.948
Technetium	Tc	43	(97)
Tellurium	Te	52	127.60
Terbium	Tb	65	158.924
Thallium	Tl	81	204.37
Thorium	Th	90	232.038
Thulium	Tm	69	168.934
Tin	Sn	50	118.69
Titanium	Ti	22	47.90
Tungsten	W	74	183.85
Uranium	U	92	238.08
Vandium	V	23	50.942
Xenon	Xe	54	131.30
Ytterbium	Yb	70	173.04
Yttrium	Y	39	88.905
Zinc	Zn	30	65.37
Zirconium	Zr	40	91.22

*A value given in parentheses denotes the mass number of the longest-lived isotope.

Emergency First Aid for Chemical Injuries

	SKIN CONTACT	SWALLOWING
ACIDS Acetic Hydrochloric Phosphoric Sulfuric	Immediately wash with plenty of running water. Remove contaminated clothing. Once all contact areas have been thoroughly washed, apply a mild alkaline solution such as soda bicarbonate. If eyes are involved, immediately flush with warm water for at least 15 min.	Do not induce vomiting. Do not give anything to an unconscious person by mouth. If conscious, wash out mouth with plenty of water, then give milk mixed with egg whites. If this is not available, give as much water as possible. Call doctor.
Hydrofluoric	Wash thoroughly and apply a magnesium oxide paste immediately. Call doctor.	Do not induce vomiting. Give plenty of water. Keep victim prone; do not move. Hospitalize immediately. Give soda bicarbonate solution immediately. Keep victim warm. Call doctor.
Chromic Acid & Dichromates	Wash with 5% sodium thiosulfate. If lesions appear, call doctor. Petrolatum in the nasal passages will protect nose from brief exposure.	
ALKALIS Ammonium Hydroxide Sodium Hydroxide Potassium Hydroxide	Wash with large quantities of water and neutralize with vinegar.	Give plenty of water mixed with lemon juice or vinegar. Follow with a spoonful of salad oil. Call doctor.
CYANIDE SALTS	Wash thoroughly. If lesions appear, call doctor.	Induce vomiting immediately. Give hydrogen peroxide mixed with water. Call doctor.
CHLORIDE COMPOUNDS Ammonium Chloride Cobalt Chloride Iron Chloride	Wash thoroughly.	Induce vomiting. Give large amounts of water. Use Epsom salts as a laxative. Call doctor.

Emergency First Aid for Chemical Injuries (contd)

ACIDS	SKIN CONTACT	SWALLOWING
CHLORIDE COMPOUNDS (contd)		
Antimony Chloride Nickel Chloride Tin Chloride	Wash thoroughly. Treat with lanolin ointment. Call doctor if irritation continues.	Give large quantities of water. Call doctor.
Cadmium Chloride		Give large quantities of water. Call doctor.
NITRATES		
Potassium Nitrate Mercuric Nitrate	Wash thoroughly. Call doctor if rash develops.	Immediately give plenty of water mixed with sodium bicarbonate, then give nonfat milk mixed with raw eggs. Call doctor.
Silver Nitrate	Wash with salt water and treat for burns.	Give 3 tablespoons of salt mixed in a pint of water. Induce vomiting.* Call doctor.
SULFATES		
Aluminum, Ammonium, Cobalt Copper, Magnesium, Nickel, Potassium, Sodium, Zinc	Wash thoroughly. Call doctor if skin reaction occurs.	Give plenty of water. Call doctor if any reactions occur.
Cadmium Sulfate	Wash thoroughly. Call doctor if skin reaction occurs.	Give plenty of water. Call doctor.

TRADEMARK CHEMICALS

Where to Buy Them

Numbers to the right of each item refer to suppliers who are given in the list of sellers directly after this list.

Chemicals not sold under a trade mark may be located in the annual *Buyers' Guide* published by Chemical Week, 330 W. 42 St., New York, N.Y., 10036, and in the Green Book published by the *Oil, Paint and Drug Reporter*, 100 Church St., New York City, 10007.

A

"ABC" 93
"A-C" 4
"ACL" 58
"ACL"-85 7
"AID-10" 31
"Accodet" 14
"Accobond" 7
"Accosoft" 14
"Acetol" 53
"Acetulan" 5
"Acryloid" 83
"Aerosil" 24A
"Aerosol" 7
"Aerotex" 7
"Aerothene" 25
"Albagel" 98
"Alcolec" 9
"Aldo" 38
"Amerchol" 5
"Amerlate" 5
"Amerscreen" 5
"Amidox" 88
"Ammonyx" 64
"Amoco" 10
"Amphoterge" 51
"Antiblaze" 19 56
"Arlacel" 45
"Armac" 13

"Armoflo" 13
"Arolon" 15
"Arquad" 15
"Atlas" 45
"Atmos" 45
"Avicel" 32A

B

"BTC" 64
"Bakelite" 93A
"Balab" 99
"Bardac" 51
"Barlox" 51
"Bayol" 31
"Be Square" 18
"Bentone" 60
"Bi-Lite" 52A
"Bio-Soft" 88
"Bioterge" 88
"Brij" 45
"Bronopol" 46A
"Bucar" 22

C

"CDB"-59 . . . . .32A
"CMC" 7 43
"Cab-O-Sil" 9A
"Calamide" 71

TRADEMARK CHEMICALS (contd.)

TRADEMARK CHEMICALS (contd.)

TRADEMARK CHEMICALS (contd.)

TRADEMARK CHEMICALS (contd.)

TRADEMARK CHEMICALS (contd.)

TRADEMARK CHEMICALS (contd.)

TRADEMARK CHEMICALS SUPPLIERS

1. Abbott Laboratories ....No. Chicago, Ill.
1A. A-D-M Co........................Minneapolis, Minn.
2. Air Products and Chemicals Inc.............Allentown, Pa.
3. Alcolac Inc........................Baltimore, Md.
4. Allied Chemical Corp................ Morristown, N.J.
5. Amerchol Corp....................... Edison, N.J.
6. American Can Co....................New York, N.Y.
7. American Cyanamid Co............... Bound Brook, N.J.
8. American Hoechst................... Somerville, N.J.
9. American Lecithin Co................. Woodside, N.Y.
10. Amoco Chemical Corp...................Chicago, Ill.
10A. Amsco Division........................Palatine, Ill.
10B. Arco Chemical Co....................Philadelphia, Pa.
11. Argus Chemical Corp.................. Brooklyn, N.Y.
12. Arizona Chemical Co....................Wayne, N.J.
13. Armak Co............................Chicago, Ill.
14. Armstrong Chemical Co..................Janesville, Wis.
15. Ashland Chemical Co.................. Columbus, O.
16. BASF Wyandotte ....Wyandotte, Mich.
17. Baker Castor Oil Co. .... Bayonne, N.J.
18. Bareco Division........................Tuisa, Okla.
19. Beatrice Foods Co. ....Chicago, Ill.
19A. Cabot Corp.......................... Boston, Mass.
19B. Carson Chemicals .... Long Beach, Cal.
20. Central Solvent & Chemical Co............. Hayward, Cal.
20a. Ciba-Geigy Corp...................... Ardsley, N.Y.
21. Clintwood Chemical Co...................Chicago, Ill.
22. Columbian Division .... Atlanta, Ga.
23. Concord Chemical Co....................Camden, N.J.
23a. Corn Products Co.................Englewood Cliffs, N.J.
23A. Croda Inc...........................New York, N.Y.
23B. Cyprus Ind-Minerals................... Los Angeles, Cal.
24. Davison Chemical Co...................Baltimore, Md.
24A. Degussa Inc......................... Teterboro, N.J.
25. Dow Chemical Co.....................Midland, Mich.
26. Dow Corning Corp....................Midland, Mich.
26A. Du Pont Co......................... Wilmington, Del.
27. Durkee Ind. Foods. ....Cleveland, O.

TRADEMARK CHEMICALS SUPPLIERS (contd.)

28.	Eastman Chemical Products Co.	Kingsport, Conn.
28A.	Emery Industries.	Cincinatti, Ohio
29.	Emlin Inc.	Kenosha, Wis.
30.	Engelhard Minerals & Chemicals.	Edison, N.J.
31.	Exxon Co.	Houston, Texas
32.	Felton Chemical Co.	Brooklyn, N.Y.
32A.	FMC Corp.	Philadelphia, Pa.
32B.	Freeport Kaolin Co.	Gordon, Ga.
33.	GAF Corp.	New York, N.Y.
34.	Geigy Industrial Chemicals	Ardsley, N.Y.
35.	General Electric Co.	Waterford, N.Y.
36.	General Mills Chemical Co.	Minneapolis, Minn.
37.	Givaudan Corp.	Clifton, N.J.
38.	Glyco Chemicals	Greenwich, Conn.
39.	Goldschmidt A. G., Th.	Essen, W. Germany
40.	Goodrich Chemical Co., B. F.	Cleveland, O.
40A.	Goodyear Tire & Rubber Co.	Akron, O.
41.	Guardian Chemical Co.	Hauppauge, N.Y.
42.	Hampshire Chemical Co.	Nashua, N.H.
42A.	Hardman Inc.	Belleville, N.J.
42A.	Henkel Inc.	Teaneck, N.J.
43.	Hercules Inc.	Wilmington, Del.
44.	Humko Sheffield Chemical Co.	Memphis, Tenn.
45.	I.C.I. America Inc.	Wilmington, Del.
45A.	IMC Corp.	Des Plaines, Ill.
45B.	ITT Rayonnico Inc.	New York, N.Y.
46.	Indusmin Ltd.	Toronto, Canada
46A.	Inolex Corp	Philadelphia, Pa.
47.	Interpolymer Corp.	Canton, Mass.
48.	Johns-Manville Co.	Denver, Colo.
48A.	Kaopolite.	Garwood, N.J.
49.	Kelco Co.	San Diego, Cal.
49A.	Lakeway Chemicals	Muskegon, Mich.
50.	Lanaetex Products Inc.	Elizabeth, N.J.
51.	Lonza Inc.	Fair Lawn, N.J.
52.	3M Corp	St. Paul, Minn.
52A.	Mallincrodt Corp.	St. Louis, Mo.
53.	Malmstrom Chemicals	Linden, N.J.

TRADEMARK CHEMICALS SUPPLIERS (contd.)

53B.	Mazer Chemicals	Gurnee, Ill.
54.	McLaughin Gormley King Co.	Minneapolis, Minn.
54A.	Mearl Chemical Corp.	Roselle Park, N.J.
54B.	Merck & Co.	Rahway, N.J.
54BB.	M. & T. Chemicals Inc.	Rahway, N.J.
55.	Miranol Chemical Co.	Irvington, N.J.
55A.	Mobay Chemical Corp.	Pittsburg, Pa.
56.	Mobil Chemical Co.	New York, N.Y.
57.	Mona Industries Inc.	Paterson, N.J.
58.	Monsanto Co.	St. Louis, Mo.
59.	Morton Chemical Co.	Chicago, Ill.
60.	N. L. Industries	Houston, Texas
61.	National Pectin Co.	Kansas City, Mo.
62.	National Starch & Chemical Corp.	Bridgewater, N.J.
62a.	Nestle Co.	White Plains, N.Y.
63.	Neville Chemical Co.	Neville Island, Pa.
63A.	Nyacol Inc.	Ashland, Mass.
64A.	Ottawa Chemical Co.	Toledo, O.
63B.	Olin Corp.	Stamford, Conn.
64.	Onyx Chemical Co.	Jersey City, N.J.
65A.	PPG Industries	Pittsburgh, Pa.
65.	P.Q. Corp.	Valley Forge, Pa.
66.	PVO International	Bridgeton, N.J.
67.	Patco Products	Kansas City, Mo.
67A.	Pennwalt Corp.	Philadelphia, Pa.
68.	Petrochemicals Co.	New York, N.Y.
68A.	Petrolite Corp.	St. Louis, Mo.
69.	Pfizer Inc.	New York, N.Y.
70.	Phillips Petroleum Co.	Borgen, Texas
71.	Pilot Chemical Co.	Santa Fe Springs, Cal.
72.	Polyesther Corp.	Southampton, N.Y.
73.	Polymer Industries	Stamford, Conn.
74.	Polymer Systems	Little Falls, N.Y.
75.	Prentiss Drug & Chemical Co.	New York, N.Y.
76.	Process Chemical Division	Morristown, N.J.
77.	Proctor Chemical Co.	Salisbury, N.C.
77A.	Quaker Oats Co.	Chicago, Ill.
78.	R.I.T.A. Chemical Corp.	Crystal Lake, Ill.

TRADEMARK CHEMICALS SUPPLIERS (contd.)

79.	Ralston-Purina Co..	St. Louis, Mo.
80.	Reheis Chemical Co.	Berkley Heights, N.J.
81.	Reichhold Chemicals Inc.	White Plains, N.Y.
82.	Rewo Chemical Co.	Bohemia, N.Y.
82aa.	Robeco	New York, N.Y.
82A.	Robinson Wagner Co..	Mamaroneck, N.Y.
83.	Rohm & Haas Co.	Philadelphia, Pa.
83A.	Sandoz Colors.	E. Hanover, N.J.
83B.	Scher Chemicals	Clifton, N.J.
84.	Shell Chemical Co..	Houston, Texas
85.	Spencer Kellogg Division.	Buffalo, N.Y.
86.	Staley Manufacturing Co., A.E.	Decatur, Ill.
87.	Stauffer Wicker Silicone Co.	Westport, Conn.
88.	Stepan Chemical Co.	Northfield, Ill.
88A.	Su Crest Corp.	Woodbridge, N.J.
89.	Sun Chemical Corp.	Cincinnati, O.
90.	Sun Oil Co.	Philadelphia, Pa.
90a.	Tanatex Chemical Co.	Wellford, S.C.
90A.	Tenneco Chemicals	Saddle Rock, N.J.
91.	Thompson Weiman Co..	Cartersville, Ga.
91A.	Titanium Pigment Co.	So. Amboy, N.J.
91B.	Tomah Products	Wilton, Wis.
91C.	Tragacanth Importing Corp.	New York, N.Y.
92.	Tretolite Division	St. Louis, Mo.
93.	Troy Chemicals.	Newark, N.J.
93A.	Union Carbide Corp.	New York, N.Y.
94.	Uniroyal Inc.	New York, N.Y.
95.	Van Dyk & Co..	Belleville, N.J.
95A.	Vanderbilt Co., R.T.	Norwalk, Conn.
95B.	Velsicol Chemical Corp.	Chicago, Ill.
96.	Vanderbilt Co., R. T.	Norwalk, Conn.
97.	Vikon Chemical Co..	Burlington, N.C.
98.	Whittaker, Clark & Daniels	So. Plainfield, N.J.
98A.	Wilson Foods Corp.	Calumet City, Ill.
99.	Witco Chemical Corp..	New York, N.Y.
100.	York Chemical Industries	Rock Hill, S.C.
101.	Ziegler Chemical Co.	Great Neck, N.Y.

"COSMETIC PLANT INSPECTIONS"

A principal feature of the Food and Drug Administration's compliance program for cosmetics is inspection of establishments engaged in the manufacture and packaging of cosmetic products. The major objective of such inspections is to determine whether products are produced under sanitary conditions, and to take regulatory action to achieve compliance. Establishment inspections afford the FDA investigator an opportunity to identify, for possible regulatory action, products which may be misbranded under the Food, Drug, and Cosmetic Act or the Fair Packaging and Labeling Act. For purposes of enforcing the FD & C Act, section 704(A) authorizes duly designated employees to enter and inspect at reasonable times and within reasonable limits any factory, warehouse, or establishment in which cosmetics are manufactured, processed, packed, or held for introduction to interstate commerce. A cosmetic shall be deemed to be adulterated "if it bears or contains any poisonous or deleterious substance which may render it injurious to users under the conditions of use". Section 601(C) states that a cosmetic shall be deemed adulterated "if it has been prepared, packed, or held under insanitary conditions whereby it may have been contaminated with filth, or whereby it may have been rendered injurious to health".

With the advent of the new cosmetic labeling regulations investigators have been instructed to report, for possible regulatory action, products that are not in compliance with the new regulations.

The fiscal 1977 establishment inspection program has incorporated a checklist on manufacturing and quality control. The checklist contains specific questions about the manner in which firms handle raw materials, labeling, manufacturing, and processing, quality control, personnel, buildings and equipment, sanitation and housekeeping. In previous years, selection of establishments to be inspected was made by the office of the Executive Director for Regional Operations (EDRO) and by the individual districts. Beginning with the fiscal year 1977, 50 percent of establishments to be inspected will have been designated by the Division of Cosmetics Technology, a proportion expected to increase to about 75 percent in fiscal 1978.

LABEL LAW IN EFFECT

The Food and Drug Administration's requirement that all cosmetics labels carry ingredient lists took effect last week amid some industry protests.

The order, upheld by a U.S. Court of Appeals last year, requires that all labels affixed to cosmetics after April 15, 1977, list in decreasing order of volume the chemicals contained in the products. Manufacturers' "trade secret" formulas do not have to be listed, except for a reference to "fragrances," "flavors" or "other ingredients."

INDEX